T0189902

NATURAL FIBER TEXTILE COMPOSITE ENGINEERING

NATURAL FIBER TEXTILE COMPOSITE ENGINEERING

Magdi El Messiry, PhD

Professor of Textile Composite Material Engineering,
Faculty of Engineering, Alexandria University, Alexandria, Egypt

APPLE
ACADEMIC
PRESS

Apple Academic Press Inc. | Apple Academic Press Inc.
3333 Mistwell Crescent | 9 Spinnaker Way
Oakville, ON L6L 0A2 Canada | Waretown, NJ 08758 USA

© 2017 by Apple Academic Press, Inc.

First issued in paperback 2021

Exclusive worldwide distribution by CRC Press, a member of Taylor & Francis Group
No claim to original U.S. Government works

ISBN 13: 978-1-77-463660-2 (pbk)
ISBN 13: 978-1-77-188554-6 (hbk)

Library and Archives Canada Cataloguing in Publication

El Messiry, Magdi, author
Natural fiber textile composite engineering / Magdi El Messiry, PhD (Professor of Textile Composite Material Engineering, Faculty of Engineering, Alexandria University, El-Gaish Rd, Egypt).
Includes bibliographical references and index.
Issued in print and electronic formats.
ISBN 978-1-77188-554-6 (hardcover).--ISBN 978-1-315-20751-3 (PDF)
1. Fibrous composites. 2. Textile fibers. I. Title.
TA418.9.C6M48 2017   620.1'97   C2016-907999-6   C2016-908000-5

Library of Congress Cataloging-in-Publication Data

Names: El Messiry, Magdi, 1942- author.
Title: Natural fiber textile composite engineering / Magdi El Messiry, PhD, author.
Description: Toronto : Apple Academic Press, 2017. | Includes bibliographical references and index.
Identifiers: LCCN 2016055286 (print) | LCCN 2016057701 (ebook) | ISBN 9781771885546 (hardcover : alk. paper) | ISBN 9781315207513 (ebook) Subjects: LCSH: Fibrous composites. | Textile fibers. | Textile fibers--Industrial applications. | Composite materials. | Polymers. Classification: LCC TA418.9.C6 N3725 2017 (print) | LCC TA418.9.C6 (ebook) | DDC 620.1/97--dc23
LC record available at https://lccn.loc.gov/2016055286

Apple Academic Press also publishes its books in a variety of electronic formats. Some content that appears in print may not be available in electronic format. For information about Apple Academic Press products, visit our website at **www.appleacademicpress.com** and the CRC Press website at **www.crcpress.com**

CONTENTS

CONTENTS

LIST OF ABBREVIATIONS

ABS	acrylonitrile butadiene styrene
AFD	average fiber diameter
BOD	biochemical oxygen demand
BOIM	biological methods
BVI	bulk volume irreducible
CAM	computer-aided manufacturing
CM	chemical method
CMC	ceramic matrix composites
COD	chemical oxygen demand
CPE	chlorinated polyethylene
CPVC	chlorinated polyvinyl chloride
CRE	constant rate of extension
CRL	constant rate of loading
CRT	constant rate of traverse
CTE	coefficient of thermal expansion
CV	coefficient of variation
DCP	dicumyl peroxide
DMA	dynamic mechanical analysis
DO	dissolved oxygen
DPF	date palm tree fiber
EBM	extrusion blow molding
EMO	environmental management system
EPDM	ethylene propylene di-monomer
EPS	expanded polystyrene
ETFE	ethyl tetra fluoro ethylene
EVA	ethylene vinyl acetate
FDM	fuse deposition modeling
FPC	fiber–plastic composites
FRC	fibers in concrete
GRP	glass-reinforced polyester
HAPs	hazardous air pollutants

HDF	high density fiber
HDPE	high density polyethylene
HIPS	high impact polystyrene
LCA	life cycle assessments
LCM	liquid composite molding
LDPE	low density polyethylene
LDPE	polyethylene low density
MACT	maximum achievable control technology
MAPP	maleic-anhydride-modified polypropylene
MCC	microcrystalline cellulose
MDF	medium density fiberboard
MF	melamine-formaldehyde
MFCs	micro fibrillated celluloses
ML	mean length
MMC	metal matrix composites
MS	margin of safety
NDT	non-destructive testing
NFPC	natural fiber composites
OFDA	optical fiber diameter analyzer
PA	polyamide
PB	polybutylene
PBAT	aromatic co-polyesters
PBS	polybutylene succinate
PBSA	aliphatic co-polyesters
PC	polycarbonate
PCL	poly-ε-caprolactones
PE	polyethylene
PEA	polyesteramides
PF	phenol-formaldehyde
PGA	polyglycolic acid
PHA	polyhydroxyalkanoates
PHB	poly-3-hydroxyalcanoates
PHB/HV	poly(3-hydroxybutyrate-hydroxyvalerate)
PHBV	poly(hydroxybutyrate-co-hydroxyvalerate)
PHCM	physico-chemical methods
PHM	physical methods

PLA	polylactic acid
PMAA	polymethacrylic acid
PMC	polymer matrix composites
PMMA	poly(methyl methacrylate)
PMPPIC	polymethylene-polyphenyl-isocyanate
PP	polypropylene
PS	polystyrene
PTFE	polytetrafluoroethylene
PU	polyurethane
PVA	poly(vinyl acetate)
PVB	poly(vinyl butyral)
PVC	poly(vinyl chloride)
PVF	poly(vinyl fluoride)
PVOH	poly(vinyl alcohol)
RF	radio frequency
RFI	resin film infusion
RIM	resin injection molding
RS	rice starch
RTM	resin transfer molding
RVI	resin vacuum infusion
SEM	scanning electron microscope
SFOC	synthetic fiber reinforcement
SLA	stereolithography
SLS	selective laser sintering
SMC	compression molding
TDS	total dissolved solids
TH	total hardness
TSS	total suspended solids
UF	urea-formaldehyde
UHML	upper half mean length
UI	uniformity index
UP	unsaturated polyester
UPVC	unplasticized polyvinyl chloride
UR	uniformity ratio
VARTM	vacuum-assisted resin transfer molding

VIP	vacuum infusion process
WCC	wood cement composites
WF	wood fiber
WPC	wood-polymer composite

PREFACE

Natural fibers have a very long practice for textile materials manufacturing. Not only for garment and household textiles, but particularly these fibers are important for the manufacture of technical textiles. Immense attention has been given to the agricultural wastes of the crops that are primarly grown for the food industry. Their waste, such as the straw, contains fibers suitable to be used in the composite fabrication. Wood and forest residuals also enhance additional sustainable resources of materials that can be converted into composites. It was established by many investigators that the mechanical properties of natural fiber polymer composites may compete with traditional glass fibers in composites. Natural fibers, as a source of raw material, not only provide a renewable supply but could also produce economic development for the countryside areas.

The impact of agricultural waste on the environment currently encourages industrial units in agricultural areas to turn agricultural waste into several products that can be utilized in different market segments, such as the automotive, electronics, sports, civil engineering, transportation, marine, wind energy, and consumer goods. As a consequence, natural fiber composites have been experiencing a healthy growth in last five years, with an annual average rate of 10% (in EU it reached 30%). In North America the yearly consumption of natural fiber composite reaches about 1.5 kg/capita. The increase of the environmental concerns generates a demand and motivation for the implementation of natural fiber composites in various new applications.

This book sheds light on the area of natural fibers composites with updated knowledge of their application, the materials used, the methods of preparation, the different types of polymers, the selection of the raw material, the elements of design the natural fiber polymer composites for a particular end use, their manufacturing techniques, and finally life cycle assessments (LCA) of NFPC.

ABOUT THE AUTHOR

Magdi El Messiry, PhD
Professor of Textile Composite Material Engineering, Faculty of Engineering, Alexandria University, Egypt, E-mail: mmessiry@yahoo.com

Magdi El Messiry, PhD, is currently Professor of Textile Composite Material Engineering in the Faculty of Engineering at Alexandria University, Egypt. He held the position of Vice Dean for Community Services and Environmental Affairs at Alexandria University from 1994 to 1998 and was the Head of the Textile Department (2000–2002 and 2005–2011). For his experience in the textile field, he was appointed as a member of the Directing Board of Spinning and Weaving Holding Company, the main Textile Company in Egypt, and is also a technical adviser to the Owner Board.

His list of publications exceeds 150 papers in the different fields of textile and materials science. He has participated in the establishment of a several textile departments in Egypt and in Arab countries and has carried out several granted projects at the international level with the colleagues from UK, France, Spain, USA, Czech Republic, Algeria, Tunisia, and Morocco. He also participates in scientific boards of several journals. In the last five years he has acted as an international expert in innovation and technology transfer. In 1999, Dr. El Messiry received the Alexandria University Award for Scientific Achievements and in 2008 the AU Achievement Award. He is the author of several books in the field of braiding, textile technology, and industrial innovation. He was a visiting professor at the Department of Materials Science and Engineering, Nonwovens Research Lab at the University of Tennessee, Knoxville, USA. Dr. El Messiry has supervised more than 40 theses for master and doctor degrees.

Dr. El Messiry was awarded a scholarship by the Egyptian government to one of the famous textile institutes in Russia (Moscow Textile University), where he obtained a PhD degree in the Design of Textile Machinery.

ACKNOWLEDGMENTS

The author wishes to thank all those who contributed so many excellent ideas and suggestions for this book. Special appreciation is extended to Mrs. N. Smirnova for her thorough review and assistance in preparing the manuscript for review and publication.

ACKNOWLEDGMENTS

DEDICATION

---

*This work is dedicated to my wife and sons for all their
endless love and support.*

INTRODUCTION

The current book is concerned with one of the most important subjects in the materials science—how to utilize natural fibers as the enforcement in composite materials that can be substituted for existing metallic and plastic parts as well as help solve the environmental problem presented by the increasing amount of agriculture residual. The book consists of eight chapters describing the nature of natural fibers, their mechanical properties and applications, as well as how to design natural fiber composites, how to choose the suitable material for a design and method of manufacturing, and finally how to test methods on the natural fibers, textile products, matrix and the composite.

The first section of the book deals with the application of textile composites in the industry and the natural fiber properties. This section not only provides an understanding of the history of natural fiber composites but also presents analyses of the different properties of natural fibers.

The second part explains textile composites, their classification, different composite manufacturing techniques, and the different pretreatment methods for natural fibers to be used in composite formation. Also, it analyzes the composite material design under different types of loading and the mechanism of failure of the natural fiber composite. The effect of the fiber volume fraction of different textile structures is highlighted.

The third part is concerned with natural fiber composite manufacturing techniques, agricultural wastes and the methods of their preparation to be used successfully in the composite, either in the form of fibers particles or nanoparticles.

The fourth part considers the testing methods of the different composite components as well as the final composite materials, giving the principle of the testing standards, either distractive or nondestructive.

This book attempts to fill the gap between the textile engineer and the designer of composites from the natural fibers. One can say that the composites material engineer should have knowledge of textile engineering in order to produce satisfactory composite for a certain application. The author has done his best to enable the reader to have broad understanding about the natural fiber composites and their end use.

—*Magdi El Messiry, PhD*

PART I

APPLICATION OF NATURAL FIBER POLYMER COMPOSITES

CHAPTER 1

APPLICATION OF TEXTILE COMPOSITES

CONTENTS

1.1 HISTORY AND EVOLUTION

The natural fibers have been used as a composite material by the ancient Egyptian. They mixed Nile mud with straw for the manufacturing of the bricks and producing stronger bricks after baking them in sun [1–3], as illustrated in Figure 1.1. The mud bricks were widely speeded technology in ancient world for building brick houses. Burning the bricks gave them more strength. It was claimed that the compression strength of such bricks may reach 19 cN/mm^2 [4]. The composition of the mud bricks consists of clay, silt, sand and straw. So, the history of consuming natural fibers in composites started at least 3000 BC. Also, natural fibers and mud formed

FIGURE 1.1   Mud bricks.

walls of homes of farmers in several *rural areas* parts of the world. Bamboo, stem fibers, wood, and other natural fibers were used for formation of mats which were utilized in the building materials. For the woven walls, as strengthen material, such as straw or hemp were used. Building of a roof involved several types of natural fibers, for instance bamboo to form a grid, covered by palm leaves to close the openings of the grid and then pasted by mud on both sides forming a multi-laminate composite.

The composite material is manufactured by several creatures. The insects teach us how to build a composite material: wasps build mud nest, paper wasps prepare their nest using wood fiber as reinforcement and their saliva as matrix forming hexagonal cells. Ovenbirds build their nests by collecting mud and manure to create a spherical shaped like then baked by the sun to create a hardened shelter [5–7].

Another example of application of Natural fiber composite was developed by the ancient Egyptians 2000 BC by inventing the papyrus. They used the inside layer of the stem of the papyrus plant, sliced it into long stripes, laid side by side forming laminate, then covered by another stripes laid 90° on the top of the first laminate, both laminates will be immersed in water and pressed together for 21 days. The plant juice, which is glue material, will bond the two laminates together forming laminated composite after being dried in sun, as seen in Figure 1.2.

Another application of the natural fiber composite is the face mask which was usually used to cover the faces of Egyptian mummies. Funerary mask, like the one shown below, were placed over the wrapped mummy as an idealized representation of the deceased. This is the example of multi laminate natural fiber composites. Figure 1.3 illustrates face mask from Ptolemaic period (300–30 BC).

FIGURE 1.2   Papyrus composite paper.

FIGURE 1.3   Face mask (Brooklyn Museum).

This mask symbolizes a complex composite material. It was made by wrapping a head shaped core (probably made from mud or straw) with layers of linen and papyrus impregnated with glue. The mask is of five-layers composite. Inner two layers were made of linen plain weave fabric, Papyrus paper on top of second linen weave covered by clay layer then the last layer which is smooth surface made of fine clay to be painted [8].

These diminutive examples of the application of composite material using natural fibers prove that the progress of the composites is not strictly modern by invention. The scientific progress over the years gives the composite applications a great evolution, especially when the new chemical resins were developed as well as the several types of manmade fibers. Referring to Figure 1.4 the landmark of composite material and its applications history. The use of natural fibers for technical composite applications has recently been the subject of the intensive research and found the enormous requests [9–26].

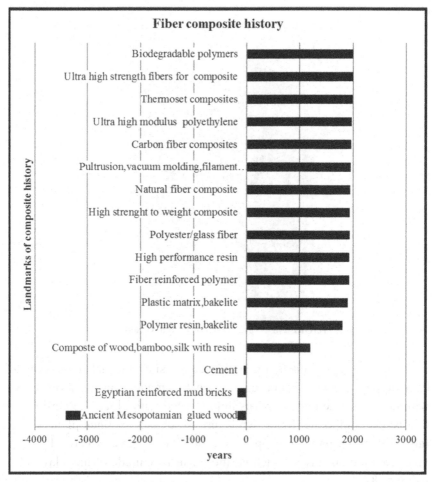

FIGURE 1.4   Composites and their applications history.

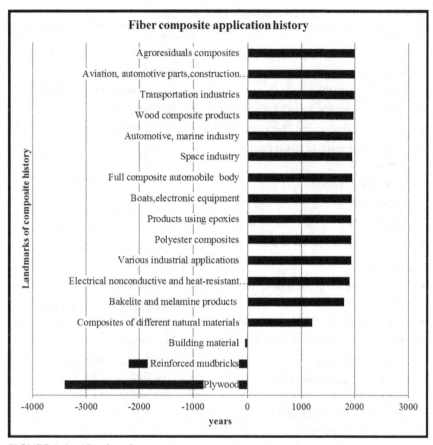

FIGURE 1.4   (Continued).

1.2   SWOT ANALYSIS OF NATURAL FIBER COMPOSITE

The SWOT analysis of the natural fiber composite identifies the four critical elements of the analysis, Strengths, Weaknesses, Opportunities and Threats. The purpose of the SWOT analysis is to assess the strength and weakness of using natural fiber composites, advantages, and the areas in which composite materials may be disapproved. Table 1.1 indicates the promising future for the natural composite applications.

Table 1.1 analysis shows the advantages of using the natural fiber composite which are: low specific weight, higher specific strength and

TABLE 1.1   SWOT Analysis of Natural Fiber Composite

Strengths	*Weaknesses*
• Increasing emphasis on recyclability	• Decrease of the world production of natural fiber
• Price-performance balance of Natural Fiber Composites	• Low production technologies
• Global concern towards global warming	• Reliability of properties
• High demand of natural fiber composites	• Design methodology
• Increasing demand of Wood Plastic Composites	• Biodegradation
• Low capital investment	
• Lower cost	
• Eco friendly	
• Renewable and sustainable plant	
Opportunities	*Threats*
• Increase demand in automotive industry by world Automotive Players	• Climate change
• Increase demand of light materials	• Shrinkage of land for natural fiber and forest
• Availability of natural fibers at low cost	• New technologies
• Change of life style	• New composite material properties requirements
• Technology development and innovation	• High performance fibers
• Increase the awareness of the environmental problems	
• New application in various industrial fields	
• High rate of composite market growth	
• New reinforcement architects are developed	

stiffness – it is a renewable resource, environmentally friendly material, and low cost. Besides, they have good thermal and acoustic insulating properties. The designers need to understand natural fiber composite weaknesses to decide what areas they should improve or use. The analysis of relative merits of natural fiber composites can also be explained by separating performance factors, as illustrated schematically in Figure 1.5 in comparison of the high performance fibers.

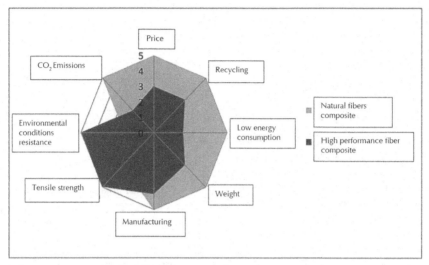

FIGURE 1.5   Natural fiber composite advantages over high performance fiber.

Natural fiber composites should not be simply applied to existing designs for other types of composites, a new design approach may be obligatory to get the best performance from natural fiber composites.

1.3   MATERIALS FOR NATURAL FIBER COMPOSITES (NFPC)

Bast fibers, such as hemp, flax, ramie, bamboo, sisal, leaf fibers, seed fibers, grass fiber, or wood fibers are suitable to be used for manufacturing of the composite materials [9–11]. Other types of fibers and fiber wastes also have found some applications as NFPC.

Not only the fibers but also their waste can be utilized. It should be mentioned that the percentage of fibers out of the total weight of the plants is varied according to the type of the natural fibers: for cotton fiber it represents only 5–10%, for flax this percentage reaches 20% after retted and 2.2–5% in tow form, hemp fibers in tow form 12% of raw hemp stalks, pineapple dry fiber gives 2.5%. Jute fiber yield 0.5 to 10.5 gm of dry fiber per plant [12–16]. This data indicates the possibilities to use the rest of the plants to manufacture composite material with different performances. Many industries manufacture composite materials reinforced with fibers

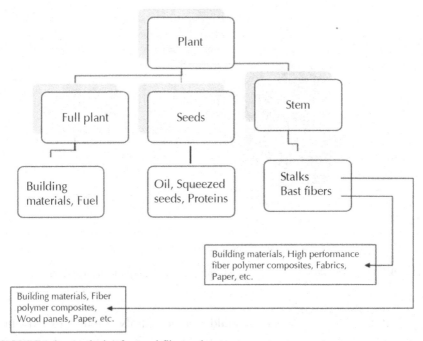

FIGURE 1.6   Analysis of natural fiber end uses.

to improve the mechanical, physical, and thermal properties of the products. Furthermore, lignocellulose materials have been treated to apply in innovative solutions for the efficient and sustainable systems. For example, the bast fiber plant products are given in Figure 1.6 which illustrates that every part of the plant can be used for various applications.

1.4   IMPORTANCE OF USE OF NATURAL FIBER COMPOSITES

The natural fibers crop reaches 25,000,000 metric tons of fibers and more than 250,000,000 metric tons of straw, besides a huge amount of wheat and rice straw as well as hard and soft wood. Most of these types of fibers are suitable to be used as reinforcement in fiber composites. The development of high performance natural fiber composites reflected on the number of articles dealing with the NFPC. The analysis of the number of articles concerned with the application of natural fibers polymer composites: (by

Google searching engine 1,780,000 publications), articles which investigating the different properties of the natural fiber polymer composites (by Google searching engine 1,070,000 publications), application of natural fiber composites in automotive industry (by Google searching engine 724,000 publications), application of natural fiber composites in aviation industry (by Google searching engine 348,000 publications), application of natural fiber composites in civil engineering (by Google searching engine 226,000 publications), design of natural fiber composites (by Google searching engine 1,840,000 publications). This review, Figure 1.7, signposts the growing interests in the application of natural fibers polymer composites in the different fields. It was also indicated that thermoplastic composite growth is greater than thermosets [17]. Many publications are dealing with the application and design of the natural fibers in the various industrial products, for example, those in which the weight of the product plays a role either in the cost, performance, or the recycling and environmental impact.

In the aerospace and automotive industries, the need to reduce fuel consumption and emissions has sparked intense interest in light weight vehicle construction. In the aerospace industry, the reduction of aircraft

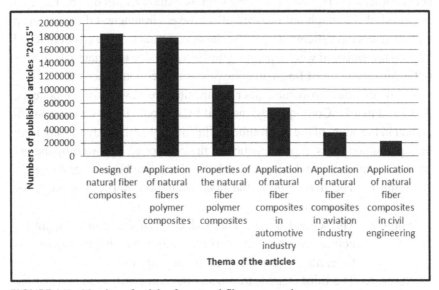

FIGURE 1.7   Number of articles for natural fibers composites.

weight is even more important than in the automotive industry to reduce fuel consumption and increase payload [18]. The benefits of using natural fiber composites result from the fact that they are made from a renewable and sustainable plant fiber source, they release no net carbon dioxide, are 40% lighter than fiberglass, and their production consumes one-fifth the energy of fiberglass production.

1.5 LIFE CYCLE ASSESSMENTS (LCA) OF NFPC

1.5.1 ELEMENT OF LCA

In the last decades, society has become concerned about the issues of natural resource depletion and environmental degradation and recycling. The environmental performance of products and processes has become a key issue, several investigations on the ways to minimize products' effects on the environment [19]. By definition the term "life cycle" refers to the major activities in the course of the product's life-span from its manufacture, use, and maintenance, to its final disposal, including the raw material acquisition required manufacturing the product [20]. The use of natural fibers will affect all the above indicators. The increase in the biomass which required to be recycled or get rid of it through burning will affect the environment. Of course, each type of fibers has different impact on the environment. More environmental impacts are enforced when extracting the fiber from the raw materials such as retting process for bast fibers, ginning process for cotton fiber.

With reference to natural fiber composite, we have two components: fibers and matrix. Consequently, the total LCA should be taken into consideration in the study of the environmental impact [19–20], firstly starting from fiber plantation to its implementation in the composite, followed by the out of service environmental impact, which may be divided into several stages:

Stage 1: The first stage is the fiber plantation, harvesting and fiber extraction.

Stage 2: The second stage is the production of the reinforcement for composite which may take the following shapes: micro particle, fibers, yarns, 2-D fabric or 3-D fabrics.

Stage 3: The third stage is to process the composite using one of the known methods.

Stage 4: The fourth stage is concerning with product in service.

Stage 5: The fifth stage is concerning with the effect of composite after service which represents recycling and disposal processes.

The second component is the matrix, which is mostly polymer or the resin as well as the coupling agent. The environmental impact of their manufacturing should be analyzed separately and added to the above impacts. Generally, transportation between different stages of the processing will increase the air pollutions through the escalations in Nitrogen dioxide, Sulfur dioxide and CO_2. The packing of the products or sub products at any stage are presenting another impact on the environment pollutions.

The LCA takes into consideration total environmental impact of the all the processes that starts with the extracting of raw materials and ends with the processing of waste and its dumping. All the phases in between, such as production, transport and maintenance, should be taken into account in the analysis. This gives the manufacturer insight in the product related environmental effects during every phase of the product's life cycle. Example of simple LCA of natural fiber composite assessments is illustrated in Figure 1.8.

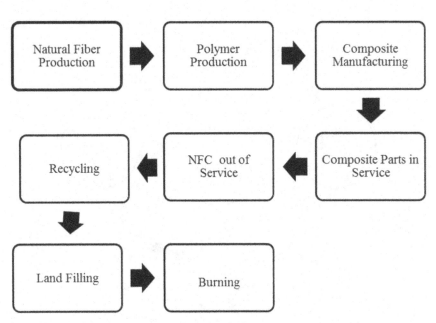

FIGURE 1.8 Life cycle of natural fiber composites.

Figure 1.9 illustrates basic flow of life cycle analysis LCA which deals with the environmental impacts at the different stages of product manufacturing from Cradle-to-grave. Cradle-to-grave is a technique to assess full range of environmental impacts associated with all the stages of a product's life from cradle to grave (i.e., from raw material extraction through materials processing, manufacture, distribution, use, repair and maintenance, and disposal or recycling) [21, 22]. The life cycle assessments are part of ISO 14000 series of Environmental Management System (EMO).

1.5.2 ENVIRONMENT IMPACTS OF NFPC

The industrial emissions may affect the environment through the increase of the water or air pollutions, as well as change of temperature and finally, increase solid waste. Any product will carry in consistence part of the pollutions element due to its manufacturing technology. Figure 1.10 illustrates impact of the product and process.

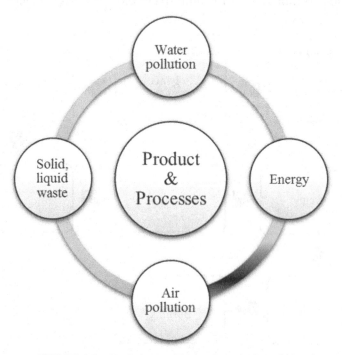

FIGURE 1.9   Basic steps of life cycle analysis (LCA).

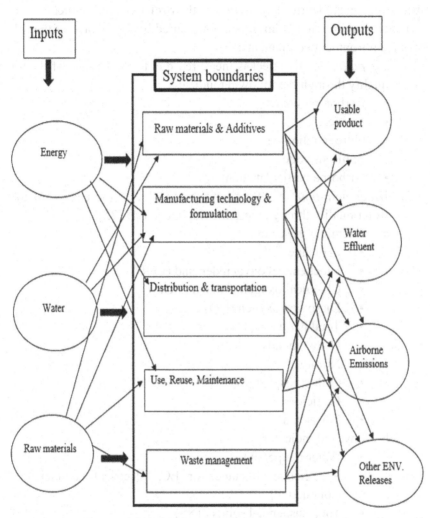

FIGURE 1.10   Impact of the product manufacturing on ecosystem.

1.5.3   ECO INDICATORS OF NATURAL FIBER COMPOSITES

Any product has an effect on the environment, which in its turn has an impact on the health, weather, and ecosystem [23]. Several researchers have been studied the LCA of the natural fiber composites [24]. Emerging comparative life cycle assessment studies of natural fiber and glass fiber composites identify key drivers of their relative environmental performance. Environmental impacts are measured by the change of the following environment components:

1. *Air pollution:* Its effect is measured due to the change of the air quality through the following indicators:
    - Carbon monoxide
    - Ozone
    - Nitrogen dioxide
    - Sulfur dioxide
    - Airborne particulate matter
2. *Water pollution:* Its effect is measured due to the change of the water quality through the following indicators [25]:
    a. Chemical indicators
        - pH value
        - Biochemical oxygen demand (BOD)
        - Chemical oxygen demand (COD)
        - Dissolved oxygen (DO)
        - Total hardness (TH)
        - Heavy metals
        - Nitrate
        - Orthophosphates
        - Pesticides
        - Surfactants
    b. Physical indicators
        - Water temperature
        - Specifics conductance or EC, Electrical conductance, Conductivity
        - Total suspended solids (TSS)
        - Transparency or Turbidity
        - Total dissolved solids (TDS)
        - Odor of water
        - Color of water
        - Taste of water

Regarding man-made fibers, LCA shows different emissions than in the case of natural fibers [26–28]. Substitution of man-made fibers with natural fibers is environmentally beneficial. Beside the environmental benefits, weight reduction is the next advantage of reduction in the weight by 20% to 50% [29, 30]. Figure 1.11 illustrates environment impact of the natural fibers, glass fibers and Polypropylene (PP) fibers [24]. The natural fibers need low energy and have low imitations. In the last decade, the trend of the bio-based polymers was recommended to replace the synthetic polymer due to their environmental advantages: Renewable raw material base, biodegradable, and lower the overall emissions and environmental impacts [31].

1.6 SOME APPLICATIONS OF NFPC

The most famous example of application of NFPC was that introduced by Henry Ford's Plastic hemp car built in 1941. Soy meal plastics were used for a steadily increasing number of automobile parts.

The composite part's formation takes place in the presence of the cellulose and other carbohydrates that were part of the soy meal (Soy meal is what is left after soy-beans are crushed or ground into flakes and the soy oil extracted with a hydrocarbon solvent). Fillers, up to 50–60%, provided

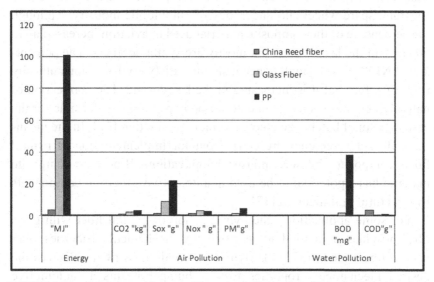

FIGURE 1.11 Environment impact of the natural fibers, glass fibers and Polypropylene fibers.

additional cellulose fibers, from hemp, wood flour or pulp from spruce or pine, cotton, flax, ramie, even wheat straw. The final mix was about 70% cellulose and 10–20% soy meal [32]. The press molding at relatively low pressures and temperatures were used in the molding process. When extra strength was required wood flour, bast fibers, cotton or wheat straw were also used.

The European and North American market for bio-fibers reinforced plastic composites reached 685,000 tons, valued at 775 million $US in 2002 [33] with an annual rate of increase of 5–10%. The consumption of wood and natural fibers composites in EU reaching the volume of Bio composite is 350,000 tons in 2012 about 25% was natural fibers [34]. At the present time, the value of 30 kg of natural and wood fiber composites is an average weight utilized per passenger cars. The number of machine parts made of natural composites reaches 30 for automotive applications. Door panels, seat backs, bolsters, load floors and packaging trays use natural fiber composites. Most natural fiber composite materials are typically manufactured from a mix of 50% natural and 50% polypropylene [35], interior trim for doors, dashboards. Wood-plastic composites are used in the interior trim for door, rear shelf, and trim for trunk/spar wheel. Cotton reinforcement composites are used in the trim for trunk/spare wheel and underbody of automobile industry. Although the percentage of the composite material used in aviation increased up to 50–55% in the last decade, but mostly are synthetic fibers. The percentage of NFPC is still small hence it should satisfy civil aviation authority rules of safety but it is increasing in the last decade. The application of natural fiber composites in an exterior body panel seems feasible for the case of a small batch size electric vehicle production [36]. In the meantime, further researches are carried out for implementation of natural fibers composites in more industrial applications. Bio-based composite material has a potential to be used in automotive structural components by structural optimization [37].

For some applications, the hybridization of natural fiber with glass fiber provides a method to improve the mechanical properties over natural fibers alone [37, 38]. Hybrid of kenaf–glass fiber enhances the desired mechanical properties for car bumper beams as automotive structural components [37]. Hybridization of kenaf fiber with synthetic fiber to form hybrid composites was manufactured using thermoplastic or thermoset matrix [39]. To improve the mechanical impact resistance, various chemical treatments, such as mercerization, silane treatment, benzoylation, etc. of bast fibers, result in better interface between fibers and matrix. The ease of processing, good mechanical properties and low overall cost have spurred on further development of bast/tannin composites [39, 40]. Even in the case of date palm tree fiber (DPF), the mechanical performance of DPF was enhanced by the different surface treatments and the chemical treatment has pronounced effect on the behavior of date palm fiber [41]. Referring to natural fiber hybrid composites (Palm fibers/Kenaf), it was proved that the hybrid composite will be more suitable for outdoor application compared to the kenaf fiber composite. The results also revealed that increasing the PET fiber content significantly improved the composite resistance to the environmental degradation [42].

Hundreds of NFPC are manufactured in the present time using natural fibers, wood residual and agricultural residual to convert them into highly demanded products [43–50]. An opportunity exists for the applications of NFPC in the development of structural beams and pedestrian bridge girders which requires low to moderate design loads. These areas where moderate strength is required and high demand is shown, offer significant opportunity where natural fiber composites can be easily introduced. The researches have shown the feasibility of the usage of natural fiber composites in various civil engineering locations including roofing and bridges [44].

The addition of 1.5% of sisal fiber changes in the performance of the fiber-matrix composite, thereby improving its toughness, increase in flexural strength and strength as compared to conventional concrete [45]. The addition of coir fibers can affect the mechanical properties of the concrete remarkably, which is highly dependent on the fiber length and volume fraction and the orientation of the fibers in concrete (FRC). Thus, a proper concrete mix design should be considered for FRC [46]. The addition of fibers in concrete structures are used in different forms for post cracking load carrying ability that increase the toughness [47, 48].

More products of NFPC are used, such as thin sheets, shingles, roof tiles, prefabricated shapes, panels, curtain walls, precast elements.

Cement bonded wood fiber formed from wood fibers in Portland cement matrix is used for the manufacturing of panel sheets bricks which are thermal, acoustical isolator, environmentally friendly material and 100% recyclable. The cement–wood composites provide a wide range of products for building applications [49]. The advantage of the NFPC makes it sustainable source ecofriendly for several products at a low cost. The total volume of 80,000 tons from different wood and natural fibers used in 150,000 tons of composites in passenger cars and lorries produced in Europe in 2012 (90,000 tons Natural Fiber Composites and 257,500 tons WPC) [50]. Wood plastic composites represent 38% of the total bio-composite used in automotive industry with hemp 25% and flax 19%. The recycled cotton and kenaf are only 7%. Recycled cotton fiber composites are mainly used for the driver cabins of trucks. In automotive industry, kenaf, hemp, abaca, and wood fibers were used for the manufacturing of door panels, seat backs, head liners, dash boards, truck liners, while in the building and construction sector bagasse, flax, coir, stalks, rise husk and wood were used for forming decks, patio, rails, wood floor panels, window frames and construction drains and pipelines. For electric industry some cases for equipment were made from kenaf fibers. The most dominant use of natural fiber composites by far can be found in interior parts of the automotive industry—other sectors such as consumer goods are still in a very early stage. In automotive, natural fibers composites have a clear focus on interior trims for doors, high value doors, and dashboard. Wood–Plastic Composites are mainly used in rear shelves and trims for trunks and spare wheels as well as in interior trims for doors. One of the most important requirements of the NFPC is the flam resistance. To improve the flammability of NFPC, the nanotechnology introduces new ideas for improving the performance of the natural fiber composite. Fire resistance of nanocomposites was evaluated with the smoke density tests. The addition of 3 % of Nano clays improved the flammability by up to 30% compared to the conventional composite, and the combination of Nano clay and bio resins doubled this value [51]. Moreover, replacing current flame retardant, such as ammonium polyphosphate (APP) by Nano clay particles is ecological and also reduces the impact of petroleum and chemical-based products. The studies ascertain that the use of well-dispersed Nano clays in polyester resin brings a comprehensive improvement and is suitable for resin infusion process. The laminates made with natural fibers showed an increase of 18% in the elastic modulus. In the present, there are several researches to solve the problems facing the use of NFPC, such as the need of increasing interfacial strength of fiber matrix in order to improve the low impact resistance of the natural fiber polymer composite and prevent fiber fragmentations [16, 40, 52].

In spite of the use of coupling agent, the flax fibers were blended with staple PLA fibers to form a homogenous fiber mixture. This enhances the delamination resistance of composite made from film stacking. The mixed fibers were converted to fiber webs using an air-laying nonwoven process, leading to isotropic composites [53]. Sugar cane bagasse-recycled polypropylene composites were successfully prepared by an injection molding technique. The increase in Young's modulus in the presence of a natural fiber is normally attributed to high modulus of the fiber compared to a thermoplastic matrix up to 20%. The elongation at break of all SB/PP composites decreases with an increase in SB content. The success of such composite is due to the effect of treatment of sugar cane bagasse by silane [54].

Another concern is the degradation of the natural fiber composites, a need of chemical treatment of the fibers. In the presence of dicumyl peroxide (DCP), the biodegradability of the composites was comparatively delayed. Depending on the end-uses of the Bio composite, we can add suitable amphiphilic additives as triggers for inducing controlled biodegradation [54], the increase of fiber value content into the composites causes acceleration of biodegradation [52]. Several technologies were used for the preparation the biodegradable composites from the natural fibers [55], such as flax/PLA bio-composites. The presence of different additives attempts to accelerate their recycles.

The following is a list of technologies or approaches having implications for the increased use of natural fibers [9, 34]: Resin Transfer Moulding, Sheet molding, Bulk molding, Extrusion and Injection, Pultrusion, Thermoforming of non-woven mats, Woven mats, Cement matrix. The film stacking, injection molding, and compression molding are the most widely used manufacturing methods. The properties of the final composites will depend on the technology of their formation which itself depends on the type of fiber used. It was found that the

TABLE 1.2 Natural Fiber/Polymer Composite (NFPC) Products and Methods of Manufacturing

Reinforcement	Matrix	Technology	Reinforcement	Matrix	Technology
Non-woven, fabrics	Thermosetting	Molding under hydraulic pressing	Natural granulated fibers	Thermoplastic resin	Melting and composting in twin-screw extruder
Natural fibers, non-woven mats, granulated natural fibers and wood flour	Polyester and thermoplastic high density or low density polyethylene	Extrusion molding, compression molding	Agriculture residues (straw, bark, fiber fines, bagasse, etc.)	Thermoplastics	Extrusion (single-screw) and compression molding
Bagasse mixed with other agricultural fibers	Thermosetting resin, phenol formaldehyde resin, methylene diphenyl disocyanate, urea-formaldehyde	Fiber de-fiberized for mat formation	Natural fibers/glass fibers	Unsaturated Polyester and Epoxy resin.	Resin transfer molding, vacuum injection
Natural fibers (jute, sisal, ramie) in mats, fabrics or hybrids	Thermosetting liquid resin, polyurethanes, epoxy resin, polyimides, polyester resins, urea-formaldehyde	Pultrusion	Natural fibers bundles, non-woven mats, bagasse	Cement and blast furnace slag mortar	Concrete manufacturing technology
Natural fibers (sisal, jute, etc.) in roving or yarn forms	Thermoplastic polymer: polypropylene, polyethylene, ethylene, propylene, rubber	Extrusions with feeders of yarns	Rice, sisal, jute, sugarcane bagasse, ramie residues	Clay	Sinterization at high temperatures
Nonwoven mat	Unsaturated Polyester resin	Resin transfer molding			

technological parameters of molding process also have the impact on the composite's final properties.

1.7 SPOT LIGHT ON MANUFACTURING TECHNOLOGY OF NFPC

There are several technologies used for the manufacturing of the natural fibers composites, which depends on the type of fibers used as reinforcement and the end use. Table 1.2 gives examples of the technologies used in the production of various types of natural fibers-based composites [10, 56, 57] and their applications. The production technique is influenced by the required fiber volume fraction for each application.

Table 1.2 point out that all the types of fibers, particles, yarns, fabrics and nonwoven materials can be used in several types of the manufacturing techniques. The suitable polymer to produce the needed parts with the predetermined specifications should be applied. In the last decades, several works were using natural fiber/glass fiber hybrid composite for the design of some parts, such as design of the automotive bumper, in order to improve the energy absorption and the impact strength as the main requirements for such structures [57–61]. In this case, the composite reinforcements may consist of the blends of fibers or laminates of individual materials chosen in a way to fulfill the essential design properties. Recently, because of the increasing environmental concern, the utilization of natural fibers from different resources, such as flax, hemp, jute, coir and sisal, etc., to replace synthetic carbon/glass fibers for reinforced polymer (FRP) composite application has gained high popularity. Natural fibers are cost effective, have high specific strength and specific stiffness and are readily available. Studies on fiber reinforced concrete indicated that short natural fibers can modify tensile and flexural strength, toughness, impact resistance. The natural fibers have potential to be used as reinforcement of concrete and/or concrete structures with good dynamic properties, increasing the energy absorption of the FCC structures [62, 63].

KEYWORDS

- **environment impacts**
- **life cycle assessments**
- **natural fiber composites**
- **natural fibers**
- **SWOT analysis**

REFERENCES

1. Aero space engineering, Composite Materials and Renewables: Wind Energy http://aerospaceengineeringblog.com/composite-materials/ (accessed January 10, 2016).
2. Pratisthan, V. V. Earth architecture, BSc Theses, Indubhai Parekh School of Architecture, Rajkote, India [Online]. http://www.slideshare.net/bhavivador/thesis-on-earth-architecture (accessed 5/4/2016).
3. Mudbrick. https://en.wikipedia.org/wiki/Mudbrick (accessed April 4, 2016).
4. Capaldi, X. Compressive Strength of Egyptian Mud Bricks https://dataplasmid.wordpress.com/category/egypt-project/ (accessed April 4, 2016).
5. Paper wasp. https://en.wikipedia.org/wiki/Paper_wasp (accessed February 4, 2016).
6. Mud dauber. https://en.wikipedia.org/wiki/Mud_dauber (accessed February 4, 2016).
7. Hornero. https://en.wikipedia.org/wiki/Hornero (accessed February 4, 2015).
8. A return to the Rubinstein cartonnage. http://www.penn.museum/sites/artifactlab/tag/cartonnage/ (accessed February 4, 2015).
9. Brouwer, W. D. Natural fiber composites in structural components: alternative applications for sisal. FAO, Economic and Social Development Department. http://www.fao.org/docrep/004/y1873e/y1873e02.htm#TopOfPage (accessed February 10, 2015).
10. Trade and Markets Division, Unlocking the commercial potential of natural fibers, market and policy analysis of non-basic food agricultural commodities team, FAO of the UN publications [Online] 2012. http://www.fao.org/search/en/?cx=018170620143701104933%3Aqq82jsfba7w&q=BAST (accessed May 1, 2016).
11. Global Materials Team FAO, New Technology for Sustainability. [Online] 2011. http://www.fao.org/fileadmin/templates/est/COMM_MARKETS_MONITORING/Jute_Hard_Fibers/Documents/Consultation_2011/3-Duarte-Ford.pdf (accessed March 10, 2015).
12. Rasheed, A.; Malik, W.; Khan, A. A.; Abdul Qayyum, N. M.; Noor, E. Genetic Evaluation of Fiber Yield and Yield Components in Fifteen Cotton (*Gossypium hirsutum*) Genotypes. IJAB. [Online] 2009, Vol. 11, No. 5, 581–585. (http://www.fspublishers.org/published_papers/27153_.pdf (accessed April 10, 2015).

13. Charlton, B.; Ehrensing, D. Fiber and oilseed flax performance. [Online] 2001. http://oregonstate.edu/dept/kbrec/sites/default/files/documents/ag/ar01chpt04.pdf (accessed February 4, 2015).
14. Lowry, G. A. Fiber flax and flax fiber. https://www.cs.arizona.edu/patterns/weaving/articles/lga_flax.pdf (accessed February 4, 2015).
15. Bast fibers. http://www.nptel.ac.in/courses/116102026/15 (accessed February 4, 2015).
16. Bos, H. L. The potential of flax fibers as reinforcement for composite materials, PhD Theses, – Eindhoven Technische Universiteit, Eindhoven, Netherlands, 2004.
17. Pereira, P. H. F.; Rosa, M.; Cioffi, M.; Benini, K.; Milanese, A.; Voorwald, H.; Mulinari, D. R. Vegetal fibers in polymeric composites: a review, Polímeros vol. 25, No. 1, São Carlos Jan–Feb 2015.
18. Preez, W.; Damm, O.; Trollip, N.; John, M. Advanced materials for application in the aerospace and automotive industries. [Online] November, 2008. https://www.researchgate.net/publication/30510916_Advanced_Materials_for_Application_in_the_Aerospace_and_Automotive_Industries (accessed March 4, 2015).
19. Mary Ann Curran, Life cycle assessment: principles and practice, EPA/600/R-06/060 [Online] May 2006. http://nepis.epa.gov/Exe/ZyPDF.cgi/P1000L86.PDF?Dockey=P1000L86.PDF (accessed March 4, 2015).
20. Sustainable packaging, design guidelines [Online] Dec. 2006. http://timrobertson.ca/innovative-tools-for-sustainable-packaging (accessed February 10, 2015).
21. Sustainable packaging, Definition of Sustainable Packaging. http://sustainablepackaging.org/uploads/Documents/Definition%20of%20Sustainable%20Packaging.pdf (accessed February 10, 2015).
22. Williams, A. S. Life cycle analysis, a step-by-step approach. ISTC Reports, Illinois sustainable technology center [Online] December 2009. http://www.istc.illinois.edu/info/library_docs/TR/TR040.pdf (accessed March 4, 2014).
23. Wotzel, K.; Wirth, R.; Flake, R. Life cycle studies on hemp fiber-reinforced components and ABS for automotive parts. Angew Makromol Chem. 1999, 272(4673), 121–127.
24. Joshia, S. V.; Drzalb, L. T.; Mohantyb, A. K.; Arorac, S. Are natural fiber composites environmentally superior to glass fiber-reinforced composites? Composites Part A 2004, 35, 371–376.
25. Water quality. [Online] https://en.wikipedia.org/wiki/Water_quality (accessed June 10, 2015).
26. La Rosa, A.D; Cozzo, G.; Latteri, A.; Mancini, G.; Recca, A. T.; Cicala, G. A Comparative Life Cycle Assessment of a Composite Component for Automotive. Chem. Eng. Transactions 2013, Vol. 32, 1723–1729.
27. Schmidt, W.; Beyer, H. Life cycle study on a natural fiber-reinforced component. Proceedings of the Total Life Cycle Conference. Graz, Austria, December 1–3, 1998, 339–348.
28. Witik, R.; Payet, J.; Michaud, V.; Ludwig, C.; Mânson, J. Assessing the life cycle costs and environmental performance of lightweight materials in automobile applications. Composites Part A 2011, 42, 1694–1709.
29. Song, Y. S.; Youn, J. R.; Gutowski, T. G. Life cycle energy analysis of fiber-reinforced composites. Composites: Part A 2009, 40, 1257–1265.

30. Begum K.; Islam M. A. Natural Fiber as a substitute to Synthetic Fiber in Polymer Composites: A Review. Res. J. Engineering Sci. [online] 2013, 2(3), 46–53 http:// www.isca.in/IJES/Archive/v2/i4/10.ISCA-RJEngS-2013-010.pdf (accessed May 15, 2015).
31. Umair, S. Environmental Impacts of Fiber Composite Materials, Master Thesis, Environmental Strategies Research, Royal Institute of Technology, Stockholm, 2006.
32. Hemp Plastic. [Online] http://www.hempplastic.com/henry.html (accessed June 10, 2015).
33. Sardashti, A. Wheat Straw-Clay-Polypropylene Hybrid Composites. MSc thesis, University of Waterloo, Ontario, Canada, 2009.
34. Bio-base news. [Online] http://news.bio-based.eu/biocomposites/ (accessed June 28, 2015).
35. Du Preez, M. J.; Damm, W. B.; Trollip, O.; John, N. G. Advanced materials for application in the aerospace and automotive industries, Science Real and Relevant. The 2nd CSIR Biennial Conference, CSIR International Convention Centre, South Africa, 2008.
36. Dammer, L.; Carus, M.; Raschka, A.; Scholz, L. Market Developments of and Opportunities for bio based products and chemicals, Final Report. Nova-Institute for Ecology and Innovation, 2013.
37. Davoodia, M. M.; Sapuana, S. M.; Ahmad, D.; Aidya, A.; Khalinab, A.; Jonoobic, M. Concept selection of car bumper beam with developed hybrid bio-composite material. Mater. Des. December, 2011, 32(10), 4857–4865.
38. Mariselvam, V.; Selvamani, S. T.; Divagar, S.; Vigneshwar, M.; Jebaraj, M.; Tensile strength evaluation on glass and sisal fiber types. JCHPS [Online] April 2015, Special Issue 9, 1–5. http://jchps.com/pdf/si9/JCHPS%20102%20V.Mariselvam%20 3–5.pdf (accessed June 10, 2015).
39. Zhu, J.; Zhu, H.; Njuguna, J.; Abhyankar, H. Recent Development of Flax Fibers and Their Reinforced Composites Based on Different Polymeric Matrices. Materials [Online] 2013, 6, 5171–5198. http://www.mdpi.com/1996–1944/6/11/5171 (accessed June 10, 2015).
40. Wang, W.; Lowe, A.; Kalyanasundaram, S. Effect of Chemical Treatments on Flax Fiber Reinforced Polypropylene Composites on Tensile and Dome Forming Behavior. Int J Mol Sci. March 17, 2015, 16(3), 6202–6216.
41. Elbadry, E. A. Agro-Residues: Surface Treatment and Characterization of Date Palm Tree Fiber as Composite Reinforcement. J. Compos. [Online] 2014, 1–8 http://www. hindawi.com/journals/jcomp/2014/189128/ (accessed May 4, 2016).
42. Abdullah, M. Z.; Dan-Mallam, Y.; Yusoff, P. S. Effect of Environmental Degradation on Mechanical Properties of Kenaf/Polyethylene Terephthalate Fiber Reinforced Polyoxymethylene Hybrid Composite. Adv. Mater. Sci. Eng. [Online] 2013, 1–8 http://www.hindawi.com/journals/amse/2013/671481/ (accessed May 4, 2016).
43. WPC for industrial applications. [Online] http://www.jeluplast.com/en/wpc/ (accessed June 15, 2015).
44. Ticoalu, A.; Aravinthan, T; Cardona, F. A review of current development in natural fiber composites for structural and infrastructure applications, Southern Region Engineering Conference, Toowoomba, Australia, 2010.

45. Chandrashekaran, B. J.; Selvan, S. Experimental Investigation of Natural Fiber-reinforced Concrete in Construction Industry. IRJET April, 2015, 2(1), 179–182.

46. Parveen, S.; Rana, S.; Fangueiro, R. Natural fiber composites for structural applications, Book of abstract, International conference on mechanics of Nano and micro composite structures, Portugal, Ferreia, A. J. M.; Carrera, E., Eds.; 2012.

47. Chowdhury, S.; Roy, S. Prospects of Low Cost Housing in India, Geo- materials. [Online] 2013, Vol. 3, No. 2, 60–65 http://dx.doi.org/10.4236/gm.2013.32008 (accessed June 30, 2015).

48. Rai, A.; Joshi, Y. P. Applications and Properties of Fiber Reinforced Concrete. [Online] Int. J. Eng. Res. Appl. 2014, Vol. 4, Issue 5 (Version 1), 123–131. http://www.ijera.com/papers/Vol4_issue5/Version%201/V4501123131.pdf (accessed Aug. 30, 2015).

49. Ronald, W.; Gjinolli, W. A. Cement-Bonded Wood Composites as an Engineering Material, The Use of Recycled Wood and Paper in Building Applications, USDA frost services and frost products [Online] 1997 http://www.fpl.fs.fed.us/documnts/pdf1997/wolfe97a.pdf (accessed June 10, 2015).

50. The timber network, New study reveals the importance and development potential of the bio-composites market. [Online] http://www.ihb.de/wood/news/Biocomposites_wood_natural_fiber_composites_nova_35056.html (accessed July 15, 2015).

51. Bensadoun, F.; Kchit, N.; Billotte, C.; Bickerton, S.; Trochu, F.; Ruiz, E. A Study of Nano clay Reinforcement of Bio composites Made by Liquid Composite Molding, Int. J. Polym. Sci. [Online] 2011, 1–10. http://www.hindawi.com/journals/ijps/2011/964193/. (May 1, 2016).

52. Alimuzzaman, S.; Gong, R. H.; Akonda, M. Impact Property of PLA/Flax Nonwoven Bio composite, Mater. Sci. [Online] 2013, 1–6. http://www.hindawi.com/archive/2013/136861/ (accessed May 5, 2016).

53. Motaung, T.E; Linganiso, L. Z.; John, M.; Anandjiwala, R. D. The Effect of Silane Treated Sugar Cane Bagasse on Mechanical, Thermal and Crystallization Studies of Recycled Polypropylene. Mater. Sci. App. [Online] 2015, 6, 724–733 http://file.scirp.org/pdf/MSA_2015080514530326.pdf (accessed May 1, 2016).

54. Kumar, R.; Yakubu, M. K.; Anandjiwala, R. D. Biodegradation of flax fiber-reinforced poly, lactic acid, express Polymer Letters. [Online] 2010, 4(7), 423–430. http://www.expresspolymlett.com/articles/EPL-0001475_article.pdf (accessed May 5, 2016).

55. UFP Technology, Molded Fiber: Environmentally Friendly Protective Packaging. [Online] http://www.ufpt.com/materials/molded-fiber.html (accessed June 10, 2015).

56. Bongarde, U. S.; Shinde, V. D. Review on natural fiber reinforcement polymer composites. IJESIT [Online] 2014, 3(2), 431–435. http://www.ijesit.com/Volume%203/Issue%202/IJESIT201402_54.pdf (accessed May 3, 2016).

57. Jeyanthi, S.; Rani, J. Influence of natural long fiber in mechanical, thermal and recycling properties of thermoplastic composites in automotive components, Int. J. Phys. Sci. [Online] 2012, 7(43), 5765–5771. http://www.academicjournals.org/journal/IJPS/article-full-text-pdf/D6B922516377 (accessed May 3, 2015).

58. Yahaya, R.; Sapuan, SM; Jawaid, M.; Leman, Z.; Zainudin, E. S. Mechanical performance of woven kenaf-Kevlar hybrid composites. *J. Reinf. Plast. Compos.* December 2014, 33, 2242–2254.

59. Kumar, P.; Belagali, S.; Bhaskar, Comparative Study of Automotive Bumper with Different Materials for Passenger and Pedestrian Safety, IOSR-JMCE Jul–Aug. 2014, 11(4), Ver. III, 60–64.
60. Umadevi, G. A.; Ramesh, N. Design and Analysis of an Automobile Bumper, IJR March 2015, 2(3), 17–22.
61. Koronis, G.; Silva, A.; Fontul, M. Green composites for an electric vehicle body: A review of adequate material's combination. Book of abstracts, 16th International Conference on Composite Structures, ICCS 16, Portugal, 2011. http://paginas.fe.up.pt/~iccs16/CD/161–200/200GKoronis.pdf (accessed May 2, 2016).
62. Banthia, N.; Bindiganavile, V.; Jones, J.; Novak, J. Fiber-reinforced concrete in precast concrete applications: Research leads to innovative products. PCIJ [Online] 2012, 57(3), 33–46. http://www.pci.org/pci_journal-2012-summer-7/ (accessed Jan 3, 2016).
63. Lu, Y. Infrastructure Corrosion and Durability – A Sustainability Study. [Online] Published by OMICS Group eBooks, 2014. http://www.esciencecentral.org/ebooks (accessed April 28, 2016).

CHAPTER 2

NATURAL FIBER PROPERTIES

CONTENTS

2.1 INTRODUCTION

The increase of the awareness by the environmental issues playing important role to encourage the industry to produce product using ecofriendly material, reducing the CO_2 emissions. The use of natural fibers to build the reinforcement of the NFPC composites aimed at reducing environmental damage, the most important property of natural fiber is biodegradability which brings it back into market [1]. The natural fiber polymer composite is also low cost and increasing the sustainability of the production. In the present time, not only the natural fibers are used for the production of the composites but also the agriculture waste found their way to produce thousands industrial products. Textile composites have the advantage of being strong and lightweight, consequently in the last decades, their demand in the automotive and aviation industries have increased. The benefits of components and products designed and produced in composite materials – instead of metals, are wanted by many industries. Natural fibers are target to be the building material of several parts: hence the substitute of metal alloy by textile composite reduces the mass by 30%. Consequently, the design of the textile composites needs the deep knowledge of the properties of their components. Bio-composite using special textile natural fibers and bio-based matrix materials in manufactures as composite of different forms are widely desired.

2.2   TEXTILE RAW MATERIALS FOR COMPOSITE

The textile composite mechanical properties depend on the mechanical properties of the fiber reinforcement, while the physical and mechanical properties of the matrix material determine the failure mode of the matrix. The fiber volume fraction decides the most of the mechanical properties. In addition to the geometric and physical properties of both fibers and matrix, the selection of fabrication of textile composites processing parameters is also very important for obtaining a textile composite with desired properties.

2.2.1   MATERIAL SELECTION

The most important problems to be solved are: choice of material, design the form and chose manufacturing process suitable to get the final shape of the designed part at the lower cost and environmental impact during the serviceability of the designed part. It's necessary to find out the optimum solution that satisfies the functional properties, manufacturing and economic problems.

The textile fibers have several properties that affect the final properties of the composite materials and their end use such as:
- Physical
  1. Moisture absorption
  2. Morphological properties
  3. Degradability
  4. Density

- Mechanical properties
  1. Strength
  2. Shear strength
  3. Ductility
  4. Young's Modulus
  5. Poisson's ratio
  6. Hardness
  7. Creep resistance
  8. Fatigue strength
- Thermal properties
  1. Thermal expansion coefficient
  2. Thermal conductivity
  3. Specific heat capacity
- Chemical properties
- Electrical properties
- Fabrication properties
  1. Ease of machining
  2. Hardening ability
  3. Formability
  4. Availability
  5. Joining techniques
- Environmental impact
- Cost

Numerous relations could help in the choice of the material such as:
- Young's modulus- density relation
- Specific Young's modulus – specific strength relation
- Thermal conductivity – expansion coefficient relation
- Natural frequency – density relation
- Strength maximum service temperature
- Strength-elongation relation
- Strength – cost relation
- Specific stiffness – specific strength relation

Several indicators express the performance index which governs material performance under specified applied loads [2, 3]. For instance, the performance index of material subjected to impact loading will depend on the strain velocity in the material which is equal to $V = (E/\rho)^{0.5}$, where E is Young's modulus and ρ is material density. Choose the performance index of material $l_1 = (E/\rho)^{0.5}$ to get the highest value. The chart of sonic velocity fiber density can be used to select the suitable material.

If during the application of a load, maximum deformation will represent the index of material performance then the choice of the material will use the strength–elongation chart of the available material. Maximum elongation percentage will divide the chart into two sections, as shown in Figure 2.1. Only four materials can be used when the maximum elongation is less than 3%.

Geometry distortion due to heat flow, the distortion $(d\varepsilon/dx)$, is minimized by selection materials with larger values of the index l_2.

l_2 = (the thermal conductivity/linear expansion coefficient).

In the case for light, stiff beam is considered. The material can be minimized by choosing material with largest value of the following index:

l_3 = (specific Young's modulus/ρ)$^{0.5}$

The material choice takes into consideration material strength, material modulus, material cost/kg. Specific strength-cost/kg and specific modulus-cost/kg are required to select the suitable material, as can be seen in Figure 2.2.

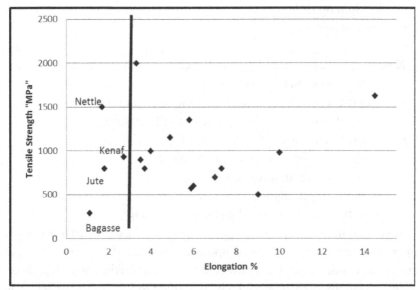

FIGURE 2.1   Tensile strength elongation chart for textile raw materials.

Fundamental steps for choice of materials are:
1. Analysis of the forces on the designed part;
2. Define of indicators that govern the performance of composite in service;

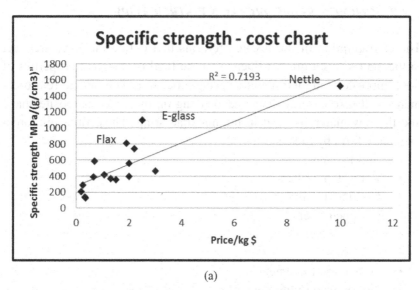

(a)

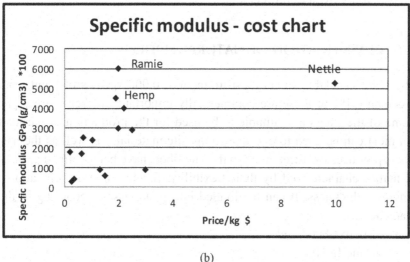

(b)

FIGURE 2.2 Specific strength-cost/kg and specific modulus-cost/kg charts of natural fibers.

3. Screening the available materials using the different properties charts;
4. Life cycle analysis.

2.2.2 CHOICE REINFORCEMENT STRUCTURE

The configuration of the textile preforms that define the fiber architect in the matrix has a great influence on the final mechanical properties of the composites and their manufacturing procedure. The fibered reinforcements will take several shapes, depending on the final composite shape and the raw materials used in the manufacturing. Thus, fiber reinforcement may form by means of:

- Short staple
- Long staple fibers
- Continuous filament
- Chopped fibers
- Micro particles
- Whiskers
- Weave reinforcement
- Non-woven reinforcement

2.3 CLASSIFICATION OF NATURAL FIBERS

There are vegetable fibers exist in about 250,000 species of plants, but less than 0.1% of these are commercially important as fiber sources [1]. Some of the fibers are suitable to be used for the production of garment but most can be used to produce composite materials, as well as the hair fibers produces by different animals. The fibers have been defined as units of matter characterized by their flexibility, fineness and a high ratio of length to thickness. It can be divided into two classes, according to the diameter size:

- Micro fibers; and
- Nano fibers.

According to the length, fibers are classified as: Nano, Micro, short, long and continuous filament. Usually, the micro fibers in diameter are used for medical and some industrial applications as well as composite materials. Textile fibers can be categorized according to its source to:

- **Natural Fiber:**
  - **Vegetable Fiber**: Cotton, Jut, Flax, Hemp, Rami, Kenaf, Sisal, Coconut Fibers, etc.
  - **Animal Fiber**: Wool, Silk, Alpaca, Camel Alpaca, Camel, Lama, Mohair, Cashmere, Angora, etc.
  - **Mineral**, Asbestos
  - **Inorganic**: Glass, Carbon
- **Man Made Fiber:**
  - **Regenerated Cellulose**: Viscose Rayon, Cellulose Acetates, Modal, Bamboo, Rubber, Lyocell
  - **Regenerated Protein**: Casein, Azlon
  - **Synthetic Fibers**: Polyester, Nylon, Acrylic, Vectran, Kevlar, Synthetic Rubber, PVC, Elastin, etc.

Here we will concentrate on the natural fibers mainly. Figure 2.3 illustrates the natural fiber classical classification. Vegetal fibers are classified

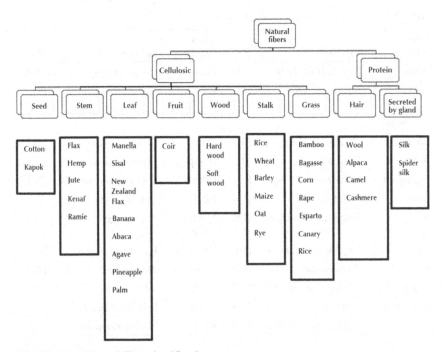

FIGURE 2.3 Natural fiber classification.

in accordance with its place of origin in a plant: (1) bast refers to fibers located in the stem; (2) leaf fibers run in the direction of the length of leaves of a plant, such as grass, and are related to hard fibers; and (3) other fibers come from the hair of seeds, mainly cotton [1].

Table 2.1 comprises the important plants of natural fibers that can be used for the manufacturing of composite materials [4].

The viability of use one or another type of fibers limited its applications in the composite industry. The global production of textile fibers increases linearly over the last decides to reach 85.6 million tons, 60.3 million tons is chemical fibers [28–31]. As illustrated in Figure 2.4a, the natural fiber production is almost constant for the last ten years. The total world production 2013 of natural fibers [34] is estimated at 33 million tons, including 26 million tons of cotton lint, 3.3 million tons of jute, 1.2 million tons of clean wool, and 900,000 tons of coir (fibers made from coconut husks). Production of all other natural fibers, including abaca, flax, hemp, kapok, ramie, sisal, silk, and other

TABLE 2.1   Important Sources of Natural Fibers

Fiber source	Species	Ref.	Fiber source	Species	Ref.
Abaca	*Musa textiles Leaf*	[5]	Coir	*Cocos nucifera*	[27]
Alfa	*Stipa tenacissima*	[6]	Cotton	*Gossypium* sp.	[9]
Bagasse (Sugarcane)		[7]	Date palm	*Phoenix dactylifera*	[10]
Bamboo	(>1,250 species)	[8]	Flax	*Linum usitatissimum*	[11]
Banana	*Musa indica*	[9]	Hemp	*Cannabis sativa*	[12]
Henequen	*Agave fourcroydes*	[13]	Kenaf	*Hibiscus cannabinus*	[18]
Isora	*Helicteres isora*	[14]	Nettle	*Urtica dioica*	[20]
Istle	*Samuela carnerosana*	[15]	Mauritius hemp	*Furcraea gigantea*	[19]
Jute	*Corchorus capsularis*	[16]	Palm Oil	*Elaeis guineensis*	[21]
Kapok	*Ceiba pentandra*	[17]	Piassava	*Attalea funifera*	[22]
Pineapple	*Ananas comosus*	[23]	Sisal	*Agave sisilana*	[1]
Phormium	*Phormium tenas*	[24]	Wood	(>10,000 species)	
Roselle	*Hibiscus sabdariffa*	[25]	Ramie	*Boehmeria nivea*	[26]

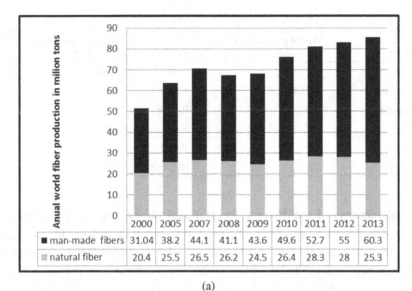

(a)

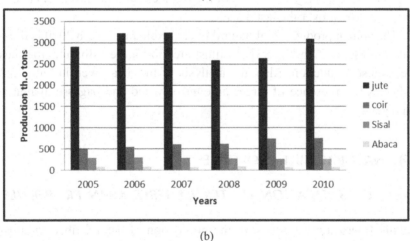

(b)

FIGURE 2.4 Production of natural fibers: (a) Total annual production of natural and manmade fibers; (b) Production of bast fibers.

fibers summed to approximately 1.6 million tons. Cotton represents the highest annual production of the natural fibers, produced worldwide throughout 50 countries, which shows the change of the production of natural and chemical fibers in the period of 2000–2013, indicating the tendency in reduction of the production of natural fibers on the contrary

TABLE 2.2   Annual Production of Natural Fibers [32, 33, 36, 51]

Fiber	World production (10³ tons)	Fiber	World production (10³ tons)
Abaca	70	Jute	2500
Bamboo	30,000	Kenaf	770
Banana	200	Ramie	100
Coir	100	Sisal	380
Flax	810	Wood	1,750,000
Hemp	215	Bagasse	75,000

to the increase of the demand for manmade fibers. The global production of fibers increased by 3.4% in 2014 to reach 96 million tons [28].

Fiber consumption in 2014 reached 93.4 million tons, representing 13.1 kg per capita. The bast fiber production is about 2 million tons [31], which represents about 10–15 % of the production of natural fibers and is almost constant over the last decades.

The annual production of natural fibers, excluding cotton 2010 is illustrated in Figure 2.4b [32, 33]. As illustrated in Figure 2.4b, jute fiber has the largest production. Also, the analysis of the data given in Table 2.2 indicates the presence of larger amount of wood and bagasse available each year.

2.4   NATURAL FIBER PROPERTIES

2.4.1   CLASSIFICATION OF TEXTILE FIBER MAIN PROPERTIES

Textile fibers are those used in the production of natural fiber composite, apparel fabrics, ropes, industrial fabrics, textile composites, protective fabrics, architect fabrics, geo-textile fabrics, medical textiles, and several other textile products. For each application, several properties should satisfy the design requirements. The main fiber properties may be classified referring to Figure 2.5.

The results of several investigations proved that each of these properties affects one or more final properties of the composite [37–39]. Table 2.3 demonstrates the interrelation between the fiber properties of

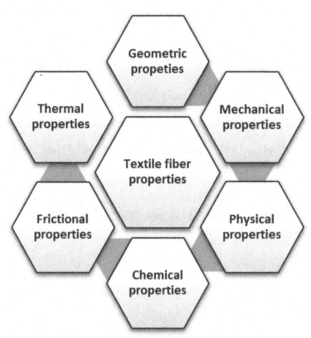

FIGURE 2.5   Classification of main textile fiber properties.

reinforcement and the final composite properties which reflects on the importance of the selection of the fibrous materials.

It can be observed from Table 2.3, that not only diameters, the stiffness and strength of fibers are the basis for the composite strength, but also the bonding properties of the fiber-matrix is important for the efficient reinforcement. The interfacial strength becomes relatively more important for natural fibers polymer composites to prevent early crack development and growth. Fiber lengths, fiber aspect ratio, fiber cross sectional shape, fiber friction, and fiber hydrophilic properties have determining effect on the values of fiber interfacial strength. One of the most important properties that affects the tensile properties of the natural fibers is moisture uptake after immersing in water – hence the specific strength and the specific modulus of most natural fibers dropped dramatically (50%) on immersion in water [40]. This is due to the hydrophilic nature of natural fiber. The chemical analysis of the natural fibers indicates that it consists of the following material elements:

TABLE 2.3   Effect of Main Textile Fiber Properties on the Properties of Composite

Composite Prop.	Mechanical properties	Bonding property	Thermal properties	Density	Fiber volume fraction %	Sound absorption	Bio-degradability
Fiber properties							
Length	X	X	X		X	X	X
Diameter	X	X	X	X	X	X	X
Bending stiffness	X	X			X		
Strength	X						X
Modulus of elasticity	X	X				X	
Density	X	X	X	X		X	X
Cross section shape	X	X	X		X	X	
Fiber crimp	X	X	X	X	X	X	
Coefficient of friction	X	X			X		
Fiber count	X	X	X	X	X	X	X
Moisture absorption	X	X		X	X		X
Fiber swelling	X	X	X	X			
Thermal properties			X				X
Sound absorption				X	X	X	
Chemical properties	X	X					X
Shear interfacial strength	X	X			X		X

- **Cellulose:** Cellulose is a natural polymer of chemical formula $(C_6H_{10}O_5)_n$ repeating units, Figure 2.6. The degree of polymerization is around 10000. Each repeating unit contains three hydroxyl groups. These hydroxyl groups and their ability to hydrogen bond play a major role in directing the crystalline packing and also govern the physical properties of cellulose [41]. Cellulose arranged in form of crystalline parts and amorphous parts, Figure 2.7. The percentage of cellulose in the fibers varied between 25% to 87 % for bamboo and cotton, respectively.

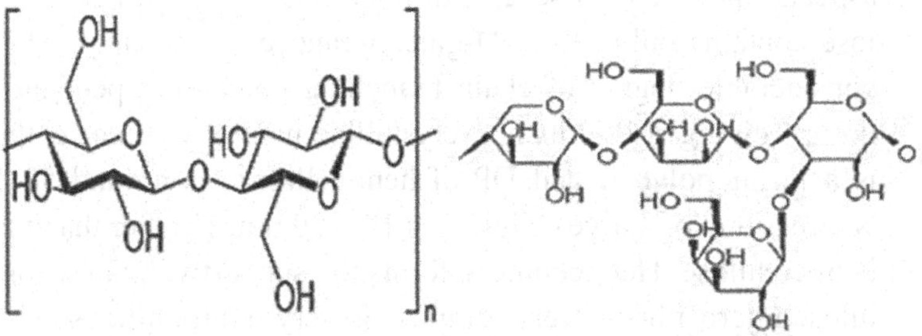

$(C_6H_{10}O_5)_n$

Cellulose   Hemicellulose

$C_9H_{10}O_2, C_{10}H_{12}O_3, C_{11}H_{14}O_4$

Lignin

FIGURE 2.6   Cellulose, Hemicellulose, and Lignin chemical formula.

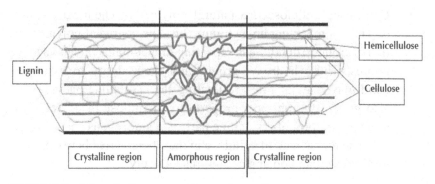

FIGURE 2.7   Micro fibrils structure of secondary wall.

- **Hemicellulose:** Hemicellulose differs from cellulose in three aspects: they contain several different sugar units whereas cellulose contains only "1,4-b-D-glucopyranose" units, they exhibit a considerable degree of chain branching containing pendant side groups giving rise to its non-crystalline nature, whereas cellulose is a linear polymer and DP of hemicellulose is around 50–300, whereas that of native cellulose is 10–100 times higher than that of hemicellulose. Hemicellulose forms the supportive matrix for cellulose micro-fibrils. Hemicellulose is very hydrophilic, soluble in alkali, and easily hydrolyzed in acids. The percentage of Hemicellulose in the fibers varied between 0.2 and 38% for Coir and Alfa, respectively [41].
- **Lignin:** Lignin is a complex hydrocarbon polymer with both aliphatic and aromatic constituents. They are totally insoluble in most solvents and cannot be broken down to monomeric units. It is the compound that gives rigidity to the plants. The percentage of Lignin in the natural fibers varied between 2 and 40% for Flax and Coir, respectively.
- **Pectin:** They provide flexibility to plants. Waxes make up the last part of fibers and they consist of different types of alcohols. The percentage of pectin in the fibers varied between 0.3 and 3% for Ramie and Abaca, respectively.

The structure of the fibers and the arrangement of the fibrils in the fiber cross-section and along the fiber length will decide the physical and mechanical properties of the fibers, even if they have the same percentage of cellulose. The structure of the fibrils is illustrated in Figure 2.7, which distinguished by the presence of two regions – crystalline region and amorphous region. The fibrils incline on the fiber axis by different angle, depending on the type of fiber [42]. The fibers shape vary between different plants but their overall shape is most often elongated with lengths in the range 1–50 mm, and diameters in the range 15–30 μm [44], the cellulose chain is having an average length of 5 μm, depending on type of species and with high percentage of lignin up to 30%. The micro fibril angle in wood fibers varies in the range 3–50°, depending on the type and location of the fibers in the wood, whereas the micro fibril angle in plant fibers is more constant in the range 6–10° [43–45].

The fibers are classified according to their lengths:
1. Short fibers (1–5 mm), (wood);
2. Long fibers (5–50 mm), (cotton, flax, hemp, jute);
3. Very long fibers >50 mm (Ramie, Coir).

For the same type of natural fibers, the variation of the properties depends greatly on the cultivation condition of the plant, soil properties, irrigation, environmental conditions and the methods of fiber extraction which are various [46]. The range of variations in some types of fibers was found very high, as illustrated in Table 2.4. For example, the strength of sisal fiber recorded to vary between 100–800 MPa. The designer of the natural fiber composite must take into consideration the variability of the fiber reinforcement properties.

2.4.2 FIBER GEOMETRICAL PROPERTIES

Generally, the fibers can be classified as staple with defined length or continuous filaments with undefined length. Most of the natural fibers have a defined length except silk, while the synthetic fibers are continuous filament and can be stapled when needed. So the first property which

TABLE 2.4 Physical Properties of Fibers for Composites [33, 44, 45, 47–53]

Fiber	Fiber length and range, mm	Average Fiber Diameter and range, mm	Density, g/cm^3	Cellulose content %
Flax	33 (9–70)	0.019 (0.005–0.038)	1.4–1.5	65–85
Hemp	25 (5–55)	0.025 (0.01–0.05)	1.48	60–77
Jute	3.5 (2–5)	0.020 (0.010–0.025)	1.3–1.5	45–63
Kenaf	2–6	0.02	1.29	45–57
Ramie	160 (60–260)	0.06 (0.040–0.080)	1.48	68.6–76.2
Sisal	1–5	0.05–0.2	1.45	50–64
Banana	–	18.5 (0.140–0.230)	1.4	63
Pineapple	3–9	6 (4–8)	1.44	81
Abaca	6.0	0.02	1.5	60
Coir	200 (100–300)	0.48	0.87–1.15	30
Cotton	37 (20–54)	0.02	1.4	85–90
Soft wood	3.3	–	1.5	40–45
Hard wood	1–2	0.022	0.6–0.9	40–50

describes the fibers is the fiber length. However, the following are the different parameters that define the fiber geometry: fiber diameter, fiber cross section shape, fiber crimp, fiber aspect ratio (l/d), fiber length, fiber fineness, and fiber cross section shape.

Each of the above parameters has a significant consequence on the fiber volume fraction, the bonding properties with the matrix and on the void percentage.

2.4.2.1 Fiber Diameter

The fiber diameter range varies from nanometer to micrometer, hence it determines the surface area per unit volume of fiber. Figure 2.8 shows the fiber diameter of the fiber denier ranges of textile fibers. The fiber diameter or thickness is one of the main properties that define the fiber volume fraction which influencing all mechanical properties of the composite. For instance, reinforcement made of yarns, the fiber diameter has additional effect on the number of the fibers in the yarn cross section and yarn diameter. The fibers are packed together with different radial packing density across the radius of the yarn.

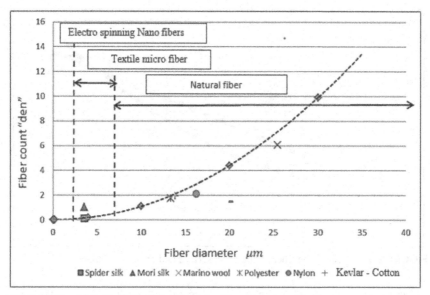

FIGURE 2.8   Classification of the fibers by diameter.

Assume that the fibers are processed from the same material but with different diameter, then the mechanical properties of the yarn will be changed, yarn tenacity, bending stiffness, creep properties, resilience, as well as bulkiness, thermal properties, water absorbance and porosity. One of the most influencing parameters is the fiber surface area which controls a number of properties, such as frictional properties between fiber, size of the pores between fibers in the yarn structure and the packing density. Figure 2.9 shows the relation between the fiber surface area and fiber cross section area.

$$d = (4 \, tex_f /\pi \, \rho_f)^{0.5} \qquad\qquad (1)$$

$$A = (\pi \, d^2)/4 \qquad\qquad (2)$$

$$S = \pi \, d \, l \qquad\qquad (3)$$

where d – fiber diameter (mm), fiber count (tex_f), ρ_f – fiber mass density (kg/m^3), A – fiber cross section area (mm^2), S – fiber surface area mm^2, and l – fiber length mm.

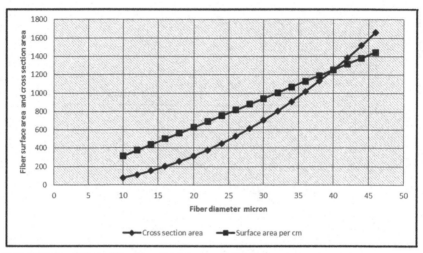

FIGURE 2.9   Effect of the fiber diameter on surface and cross section areas.

The cross section area affects most of the mechanical properties, particularly the interfacial properties of the fiber matrix.

Interfacial shear strength at the fiber-matrix interface τ was determined by means of a fragmentation test on single fiber composites. Assuming the strength of fiber matrix interface τ, l – fiber length, d – fiber diameter, and σ – fiber stress, then fiber–matrix interfacial strength τ will be:

$$\tau = \sigma \, (d/2l) \tag{4}$$

This indicates the dependence of the fiber interfacial force on the fiber aspect ratio and the fiber strength. Consequently, the mechanical properties will alter by the change of yarn diameter, such as strength, tear strength, shear property. This will directly reflect on the natural fiber composite mechanical properties. In the natural fibers, the diameter is not constant but depends on several factors related to the fiber growth and the time of harvesting and methods of fiber preparation technology [54]. Figure 2.10(a–e) show the frequency distribution of fiber diameter of some fibers. In the bast fibers, the diameter of the fibers depends on the pretreatment process of the fibers. Figure 2.10(a–d) show the frequency distribution of some types of bast fibers, which indicated that it is a binominal distribution with skewness to the right [54]. The coefficient of variation of the

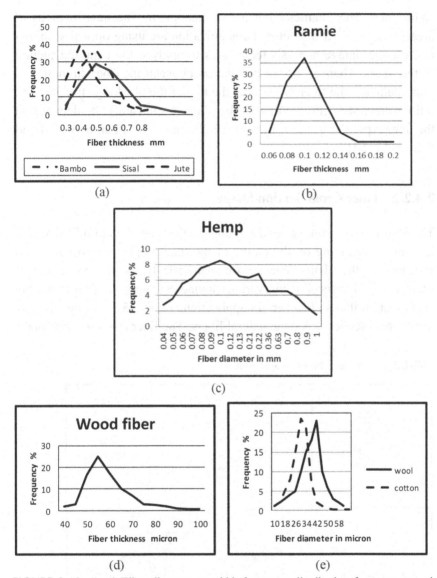

FIGURE 2.10 (a–e) Fiber diameter or width frequency distribution for some natural fibers.

fiber thickness varies between 20–35%, which indicates the sources of isotropic properties of the natural fiber composites (NFPC).

For fiber diameter measurement and distribution, there are several methods, the simplest one is the airflow method which can give the mean some fibers fiber diameter but not its histogram. The new methods of measuring the fiber diameter characterization are using optical scanning microscope, image analysis techniques to analyze the captured picture of the fibers. There is a high correlation between the fiber strength and fiber diameter, the finer is the fiber the higher the breaking strength. The ratio of fiber strength (MPa) and fiber diameter in (mm) "β" depends on the fiber type. βsisal = 39, βjute = 19, βramie = 23, βbamboo = 54 and βcior = 13 [54].

2.4.2.2   Fiber Cross Section Shape

The fiber cross section varies depending on the type of natural fiber which is formed depending on the nature of the fiber and to perform a certain function for the plant. Whereas for manmade fibers, the cross section shape can be produced using various spinneret shapes in order to get the cross section that suits a certain application. Table 2.5 illustrates the different cross sections of some natural fibers. The fiber cross sections plays

TABLE 2.5   Natural Fibers Cross Section

| Cotton | Wool | Mohair |
| Flax | Hemp | Silk |

a pronounced role in determining the behavior of fibers in yarn as well as the frictional property of the fibers, which is one of the main factors that strengthen the yarns. The fiber bending stiffness will depend on the shape of the fiber cross section, therefore it is expected that the fiber matrix interfacial strength will be affected.

These shapes of fiber cross section affect the following fiber properties:
- Fiber surface area
- Fiber bending stiffness
- Fiber tenacity
- Fiber water absorption
- Fiber frictional properties
- Fiber interfacial strength

Consequently in their turn, these properties will have the impact on the composite failure mechanism.

2.4.2.3 Fiber Length

As was mentioned, the most popular natural fibers for composites are bast fibers isolated from the stem by several processes, breaking, scotching and hackling. The fibers are packed together in tape like form. Figure 2.11 illustrates how the fibers are constructed [55–57].

It can also be seen how the technical fibers consist of elementary fibers which have lengths between up to 5 cm, and diameters between 5 and 35 μm. The technical fiber consists of about 10–40 elementary fibers in cross section. The elementary fibers overlap over a considerable length and are glued together by pectin and hemicellulose [56]. The fiber length and thickness can be changed through pretreatment processes that will define the fineness and length of the technical fiber. This explains the wide range of the physical and mechanical properties of the bast fibers.

Fiber length of natural fibers is given in Figure 2.12 with its shortest value for wood fibers and longest for ramie and coir fibers. Bast fibers can be cut into shorter fiber, according to the designer requirements. The short fibers are easier to deal with in the cases of the fiber plastic composite formation by extrusion. The fiber length should be higher than the critical length in order to insure that the fibers will fail under stress equal to their tensile stress.

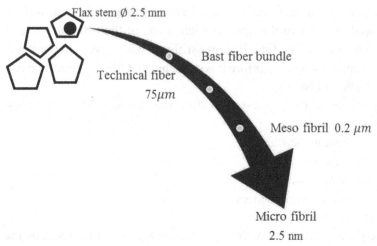

FIGURE 2.11   Flax stems components.

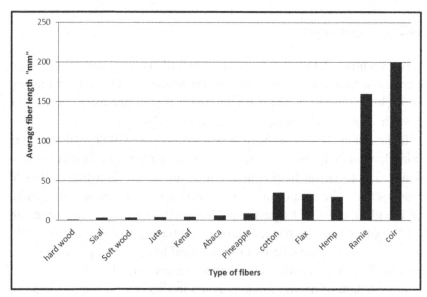

FIGURE 2.12   Fiber length of some natural fibers.

2.4.2.4   Fiber Aspect Ratio

Fiber aspect ratio is the ratio between the fiber length and fiber width (fiber diameter). It affects the distribution of the shear stress along the fiber length [37], which determines the critical fiber length. It was found that the tensile strength trends to increase with the increase of the fiber length but the elongation at break of the composite was not affected significantly [58]. Table 2.6 gives the aspect ratio of some natural fibers. For manmade fiber, it is possible to produce fibers with the wanted aspect ratio that satisfies the end use of the fabric. Consequently, it is potential to manufacture composites with different types of fibers of long length or the short. Long fibers can be cut to shorter length to suite the limits of the composite design. The fine fiber gives greater number of fibers per unit volume, whiskers or micro particles of different fibers can be used as filling material in some types of composites. The finer is the diameter, the higher will be total surface area. This will influence all the mechanical properties of the composite. The fineness of the bast fibers can be changed through chemical treatment of the fibers, such as caustic soda treatment, which leads to debonding of the micro fibril and reducing the fiber diameter. The smaller fiber diameter will affect negatively the manufacturing of the composites due to the problem of the fiber aggregation which prevents the even distribution of the fibers in the composite mass, increasing its heterogeneity. Fiber volume fraction increases as aspect ratio increases,

TABLE 2.6 Fiber Aspect Ratio [44, 49–53]

Fiber type	Fiber length, mm	Fiber fineness, mm average (range)	Fiber aspect ratio
Flax (single fibers)	33 (9–70)	0.019 (0.005–0.038)	1738
Hemp (single fibers)	25 (5–55)	0.025 (0.01–0.05)	1000
Jute (single fibers)	(2–5)	0.020 (0.010–0.025)	100
Cotton	10–60	0.02	900–1650
Kenaf	2–6	0.2	20
Ramie	60–260	0.040–0.080	1500
Sisal	1–5	0.05–0.2	20
Bamboo	2.7	14	200
Straw	1.4		50
Wool	150	0.02	5000
Soft wood	3.3		100
Hard wood	1		50

leading to have fiber reinforcement of low stiffness and higher strength. From Table 2.6, wool, bast fibers and cotton have the highest fiber aspect ratio. If the aspect ratio of the fiber is lower than the critical aspect ratio, insufficient stress will be transferred and the reinforcement will be inefficient. The suggested aspect ratio is in the range of 40–220 for short fiber composite [59].

2.4.2.5 Fiber Crimp

The fiber crimp or waviness is natural shape of the fiber or is imposed due to the different pretreatment processes [60]. The crimp in textile fibers affects the voluminosity of the fiber, influencing the strength of the composite [61]. The specific density of the yarns and fabric is a function of the crimp presence. Crimp in textile fibers is defined as waves or curls in the strand, induced either naturally during fiber growth, mechanically, or chemically. Fiber waviness is fluctuation in the fiber orientation that is inherent in the sliver morphology of plant-based natural fibers, it proved that negative correlation between composite strength and the fiber weave [62]. Fiber crimp is expressed as waves or crimps per unit length as presented in Figure 2.13. The crimp in fibers may be in form of convolutions, as cotton fibers, waves as in wool fiber, or planer crimp as mechanically imposed in manmade fibers. The crimp may be two-dimensional or three dimensional.

The quantified parameter expressing the degree of the fluctuation is called the "area ratio, [62]. The amount of crimp corresponds to the fine-

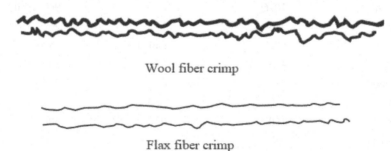

Wool fiber crimp

Flax fiber crimp

FIGURE 2.13   Wool and flax fiber's crimp.

ness of the fibers. Fine wool, like Merino, may have up to 40 crimps per cm. The cotton fiber's convolution, which gives helical shape to the fiber, is different depending on the cotton variety. It has 1.2–1.6 convolution per cm. Cotton varieties with small convolution angle have higher fiber tenacity.

The fiber convolutions does not always have a definite helical path as illustrated in Figure 2.14, but it may change its direction or flatten depending on the maturity of the fibers [63]. There are several terms related to crimp; crimp frequency, the number of full waves in length of fiber divided by the straitened length, crimp contraction, percentage of contraction of the fiber length due to its crimping, and crimp angle. Figure 2.15a shows idealized geometry of crimped fiber that defines the different parameters of the crimp.

$$\text{Crimp frequency} = 4l_0/\lambda \tag{5}$$

$$\text{Crimp angle} = \Phi \tag{6}$$

$$\text{Crimp contraction or crimp index} = ((2l_0 - l)/2l_0) \times 100 \tag{7}$$

Degree of crimp $= 2l_0/l$; and Crimp depth (is the perpendicular distance between a peak of a crimp and a line joining the valleys of the adjacent crimp waves) = 2A. Here, L_0 = side length of one crimp bow, l = width of one crimp bow, λ = wave length of crimped fiber, Φ = crimp angle, a crimped amplitude.

For manmade fiber, the crimp frequency is ranging from 4 to 23. The fiber crimp may have helical path, as presented in Figure 2.15b.

FIGURE 2.14 Cotton fiber convolutions.

FIGURE 2.15   Crimp structure.

2.4.3 MECHANICAL PROPERTIES OF TEXTILE FIBERS

The strength of the composite material depends mainly on the strength of the reinforcement which is a function of the strength of the fibers. The fiber composite materials are subjected to the forces during its life which are different in magnitude and direction, also they may be of constant value or time dependent. It is important to calculate those forces and how materials are deformed or failed as a function of applied load. It should be mentioned that the textile material is a viscoelastic material so as the polymer of the matrix of different properties. These incompatibilities of the main two elements of the composite define the mode of failure of the composite material. The fiber mechanical properties expressed by:

1. Breaking load "N" or tenacity (g/denier, gm/*tex*, N/*tex*, or MPa);
2. Breaking elongation percentage;
3. Work of rupture: is defined as the energy required breaking a material or total work done to break that material (Unit: Joule (J/m^2);
4. Initial modulus, modulus of elasticity;
5. Work factor;
6. Elastic recovery: the percentage of the recovered elongation after the removal of load;
7. Creep: the behavior of the fiber when subjected to a constant load as a function of time.

Table 2.7 gives the mechanical properties of the most used fibers for composite manufacturing. It should be noticed from the analysis of Table 2.7 that the mechanical properties of the natural fibers have high variability and must be considered during the design of the natural fiber polymer composite.

It returns to the point that the properties of the fibers widely vary, depending on the location where the fibers were taken along the stem of

TABLE 2.7 Physical and Tensile Properties of Natural and Glass Fibers Used for NFPC [33, 44–53, 64]

Fiber	Diameter (μm)	Density (g/cm³)	Tensile strength (MPa)	Elongation (%)	Elastic modulus (GPa)	Specific modulus (GPa/(g/cm³)	Specific strength (MPa/(g/cm³)
E-glass	<17	2.5–2.6	2000–3500	1.8–4.8	70–76	29	1100
Abaca	122	1.5	400–980	1.0–10	6.2–20	9	467
Alfa	5–20	0.89	1350	5.8	22	25	1517
Bagasse	10–34	1.25	222–290	1.1	17–27.1	18	204
Bamboo	25–40	0.6–1.1	140–800	2.5–3.7	11–32	25	588
Banana	12–30	1.35	500	1.5–9	12	9	370
Coir	10–460	1.15–1.46	95–230	15–51.4	2.8–6	4	127
Cotton	10–45	1.5–1.6	287–800	3–10	5.5–12.6	6	355
Curaua	7–10	1.4	87–1150	1.3–4.9	11.8–96	39	714
Flax	12–600	1.4–1.5	343–2000	1.2–3.3	27.6–103	45	807
Hemp	25–600	1.4–1.5	270–900	1–3.5	23.5–90	40	745
Henequen	340	1.2	430–570	3.7–5.9	10.1–16.3	11	417
Isora	10	1.2–1.3	500–600	5–6	10	8	440

TABLE 2.7 (Continued).

Fiber	Diameter (μm)	Density (g/cm³)	Tensile strength (MPa)	Elongation (%)	Elastic modulus (GPa)	Specific modulus (GPa/(g/cm³)	Specific strength (MPa/(g/cm³)
Jute	20–200	1.3–1.49	320–800	1–1.8	30	30	400
Kenaf	40–81	1.4	223–930	1.5–2.7	14.5–53	24	414
Nettle	19–50	0.72	650–1500	1.7	38	52.8	1527
Oil palm	250–610	0.7–1.55	150–500	17–25	80–248	0.5–3.2	287
Piassava	1100	1.1–1.45	134–143	7.8–21.9	1.07–4.59	2	96
PALF	20–80	0.8–1.6	180–1627	1.6–14.5	1.44–82.5	35	750
Ramie	20–80	1.0–1.55	400–1000	1.2–4.0	24.5–128	60	560
Sisal	100–400	1.33–1.5	363–700	2.0–7.0	9.0–38	17	393
Coconut	12–25	1.1	131–175	30	4.0–6.0		1363

the plant. The basic values required for the design of the natural fiber polymer composites are:

- Density, (g/cm³);
- Tensile strength, (MPa);
- Elastic modulus, (GPa);
- Specific modulus, (GPa×cm³/g);
- Elongation at failure (%);
- Fiber length.

A comparison between the different properties of fibers, which can be used as reinforcement of the natural fiber polymer composites, are demonstrated in Figure 2.16.

The values of fiber properties are essential for the designer to decide their suitability for various end uses. The method of composite manufacturing depends also on every fiber physical, mechanical, thermal, and even electrical properties. Fibers able to carry out electrostatic charges will require further attention to prevent the process disturbance. Generally, there are minimum values for the various properties, such as:

- Tenacity > (30 cN/tex);
- Extensibility (less than 3%);
- Fiber aspect (Length to width ratio), (1000–2500);
- High modulus of rigidity (80 cN/denier or more conditioned, 50 cN/denier wet);
- Cohesiveness (low value);
- Uniformity (high);
- Crimp (low crimp);
- Fineness (9–50 den);
- Fiber coefficient of friction (μ 0.2–0.3);
- Low swelling in water;
- Low water absorption.

Reliant on the value for each of the above properties is hinge to the final properties of the designed matrix.

2.4.3.1 Fiber Strength

The composites are subjected to various types of loading during it service life. These forces differ depending on the type of processing operations,

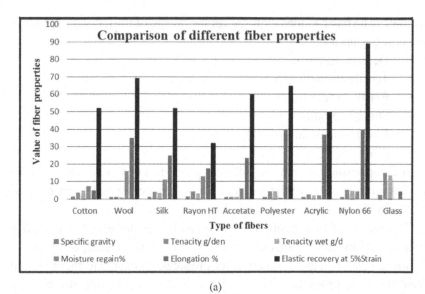

(a)

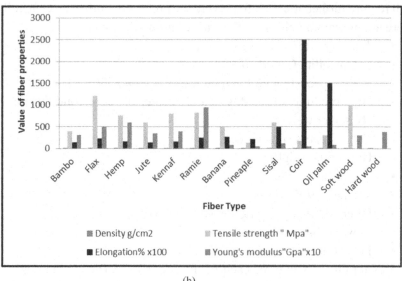

(b)

FIGURE 2.16 Properties of textile fibers.

design of the composite, and its end use. Consequently, the choice of fiber mechanical properties is highly imperative. The stress strain curve of cotton fibers, shown in Figure 2.17, illustrates the variability of the strength properties. The entire bast fiber's strength elongation curve is almost straight line. Figure 2.18 shows a comparison of stress strain curves of some other types of natural fibers.

The strength and strain of the different natural fibers are compared in Figure 2.19, demonstrating that Alfa fiber has the largest value of strength and Coir fiber the highest strain at break, and most of the fibers have high value of Young's modulus.

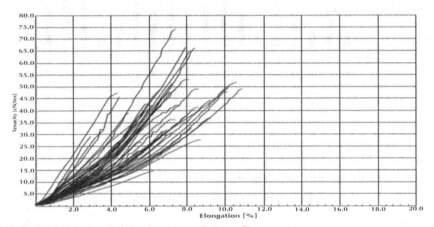

FIGURE 2.17 Load elongation curve of cotton fiber.

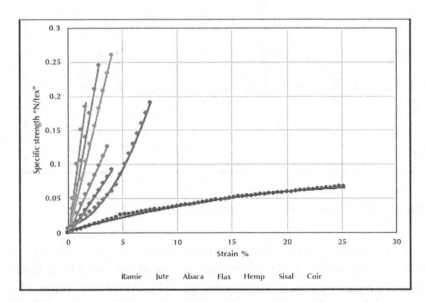

FIGURE 2.18 Specific strength-strain % curve of some natural fibers.

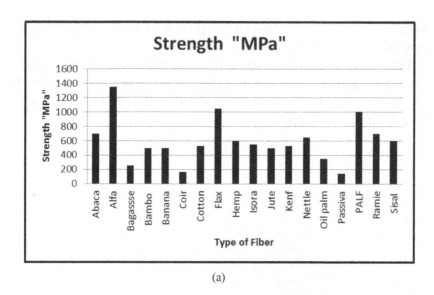

(a)

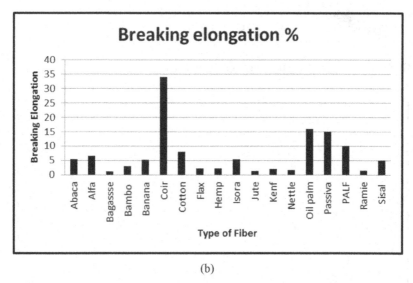

(b)

FIGURE 2.19   Strength and breaking elongation % of natural fibers.

The Alfa fibers have the highest strength and reasonable elongation while Coir fibers have highest elongation and low strength. Oil Palm fibers show the moderate values of both strength and breaking elongation. The values of the specific strength and stiffness of the fibers are interrelated with the mechanical properties of the fibers by its weight which is one of the most advantageous property of using natural fiber in the composites.

Specific strength of the natural fibers not only dependent on the percentage of cellulose in the fiber but mostly on the fiber structure. The specific modulus is also a function of the fiber density, as interpret in Figure 2.20. Both properties are a function of the fiber structure, the percentage of amorphous to crystalline, as well as the fibril size and its lay-out in the fiber. The ranking of the fiber is different when considering specific strength or specific modulus of elasticity, Figure 2.21. Ramie or Alfa fibers are good examples.

Figure 2.22 illustrates the positive correlation between the specific strength and the specific modulus of elasticity, coefficient of correlation 0.87 (highly significant).

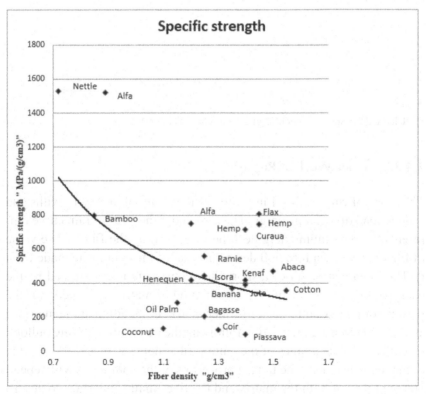

FIGURE 2.20   Specific strength versus density of the fibers.

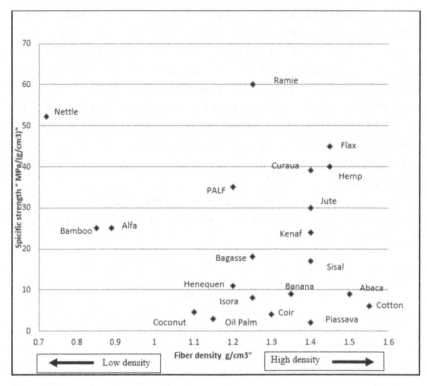

FIGURE 2.21   Specific modulus of elasticity versus fiber density.

2.4.3.2   Fiber Work of Rupture

The work of rupture is an indicator of the material ability to withstand sudden load of a given energy. The area under the stress strain curve represents the maximum energy can be withstood by the fibers. The value of fiber work of rupture will determine that of the composite made from it. The impact strength of a short fiber composite is determined by the energy required for the plastic deformation of matrix and fibers, for the fracture of matrix and fibers, for fiber-matrix de-bonding at the interface and for overcoming the friction following the de-bonding during pullout. One of the disadvantages of the natural fiber composite is the low value of impact resistance. The impact properties of the composite will depend on the ability of both the matrix and the fiber-matrix interface to absorb

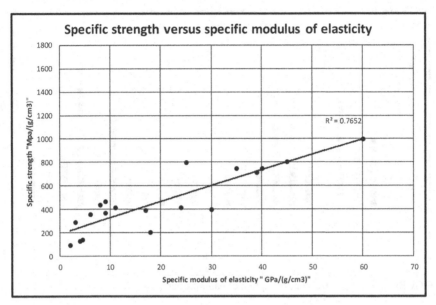

FIGURE 2.22 Specific strength versus modulus of elasticity for natural fibers.

the high-speed impact. The quality of the fiber-matrix interface is significant for the application of natural fibers as reinforcement fibers for polymer matrices. The impact strength of natural fiber-reinforced composite
can be considerably changed with the alteration of fiber-matrix adhesion
by using different types and amounts of coupling agents. The impact
strength can be increased by providing flexible interphase regions in the
composite or by using impact modifiers. The use of an impact copolymer
improves the impact resistance [65] with some reduction in modulus
and strength of the composite. Natural fiber polymer composites have
excellent mechanical properties as high specific strength and modulus
however low impact load because of low interfacial strength between
the fibers and the polymer [66]. This does not ignore the necessity to
use the fibers possessing high value of work of rupture for more energy
absorption.

Classification of natural fibers according to their specific work of rupture indicates that PALF fibers have highest value while the bast fibers
have moderate values, Figure 2.23.

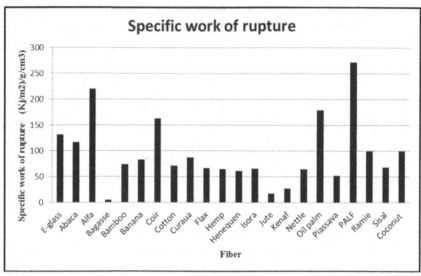

FIGURE 2.23   Specific work of rupture of natural fibers.

2.4.3.3   Fiber Creep

When the textile material is subjected to stress, the material is instantaneously deformed and will continue to deform under the same value of stress: with the time the strain in creep will take the shape as illustrated in Figure 2.24. Creep strain starts with instantaneous deformation – to primer deformation – to secondary deformation and ended by a zone which characterized by high rate of deformation completed by rupture. When the stress removed, the material will recover its deformation in the same order, ending by the permanent deformation. The creep property is very important for textile material application. The values of instantaneous deformation, secondary deformation, instantaneous recovery, and permanent deformation characterize the creep phenomena. All these values will vary according to the temperature surrounding the material and the applied load.

Creep strain can be described by the simple formula:

$$\varepsilon_c = a\,\sigma^n\,t^b \tag{8}$$

where: a, n, and b are constants dependent on the temperature, ε_c is the creep strain and σ is the applied stress. The above equation was applied

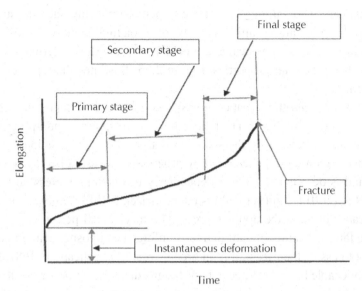

FIGURE 2.24 Creep strain vs. time.

to the experimental creep data at room temperature. The increasing interest is the creep properties of the fiber-reinforced composite due to the fact that the variable nature of loading applied on the composites during their services. Generally, the creep in thermoplastic materials (matrix) involves the rupture of the interfacial bonding under the applied stress. The increase in the surrounding temperature will increase this phenomenon. The creep behavior of the composite can be improved by the use of the natural fibers that possess a low value of instantaneous deformation or secondary creep zone, including the use to the elevated temperatures [67]. The matrix, which is a viscoelastic, under a constant load will creep and the strain will transfer to the fibers, thus damage the interfacial bonding. The difference in the creep property of the matrix and reinforcement will lead to the interfacial de-bonding propagation and promote composite failure. The creep of bast fiber has low instantaneous strain values on the contrary to polymer.

2.4.3.4 Fatigue Properties of Natural Fibers

The majority of components of a design involves parts subjected to fluctuating or cyclic loads. Most of structural failures occur through a fatigue mechanism. The fatigue sensitivity depends on many factors, such as material type, material uniformity, type of loading cycle, value of maximum fatigue stress, physical and mechanical properties, internal structure of the fiber, thermal properties. Figure 2.25 illustrates the types of basic cyclic loading.

On semi-logarithmic paper, the S-N curve becomes approximately trilinear, Figure 2.26. Region I ("static region") corresponds to non-progressive fiber damage, defined by the static strength of the fibers. Region II ("progressive region") is characterized by progressive surface fiber micro cracking, and is described by linear dependency of maximum stress on the log (N). Region III ("limit region") is characterized by fibers internal structure that can withstand the applied stress. The mechanical performance under cyclic fatigue loading of natural fiber-reinforced composites is an important aspect to assess their feasibility in broad structural applications. [68]. There is a noticeable lack of fatigue data of fatigue data of the fibers as well as the natural fiber composites. The fiber static properties have insignificant effect on the fatigue sensitivity of the NFPC slope of S-N curve [69].

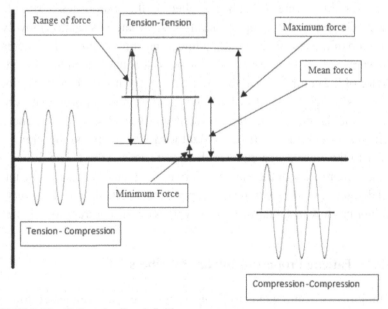

FIGURE 2.25   Fatigue loading definitions.

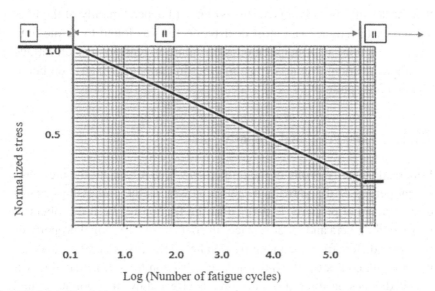

FIGURE 2.26 S-N curve.

The S-N fatigue lifetime curves of the fibers are presented in Figure 2.26, Wohler equation:

$$\log(N_f) = a - b\,\sigma_{max} \tag{9}$$

where: N_f is the number of cycles at rupture, a, b are material parameters, and σ_{max} is the applied maximum stress. The value of b depends on the type of fiber and the pretreatment processes.

The fiber fatigue performance in NFPC was addressed by several investigators [70], normally in all cases, the normalized stress decreases after the fibers subjected to the fatigue cycles. The higher the loading level $r =$ (the maximum displacement during the sinusoidal fatigue loading/the maximum displacement of the specimen at failure), the larger will be the hysteresis area and the higher the exponential term involved [71]. Experimental results clearly show that the number of cycles to failure at a given loading level increases as the loading level decreased in fiber. The stiffness of the fibers slightly increases in most bast fibers. Chemical pre-treatment of fibers gives a significant assistance to the fatigue life of the fibers as well as to natural fiber-polymer matrix composites. It should be mentioned that the sensitivity to fatigue of the NFPC will depend mainly of the fiber sensitivity to fatigue. The most decisive factor was found to be the fiber volume fraction, as principal modes of failure in tension–tension fatigue are a combination of matrix fracture, fiber fracture, fiber/matrix de-bonding, and fiber pullout [72–74].

2.4.3.5   Fiber Frictional Property

Frictional property is one of the properties which affects all the mechanical properties of the yarns and fabrics as well as the processing parameters. Frictional property is defined as the friction between the fibers or between the fibers and other surfaces in contact. The frictional property of the fibers differs from the metal friction hence the fibers in most cases are not straight, consequently under no load a friction force is actuated which is called cohesion force. Friction force is affected by fiber count, normal force applied, roughness of the fiber surface and area of contact, fiber crimp, and a cross sectional shape. Because of the complex nature of fiber friction, the friction coefficient value has been found to be strongly reliant on testing conditions. The coefficient of friction between fibers depends on several parameters, such as material properties, surface morphology, shape of fiber cross section, crimp, relative humidity, presence of natural or added lubrication materials. The kinetic coefficient of friction is less than the static one. Figures 2.27 gives the coefficient of friction of some fibers [75]. The degree of roughness of the fiber surface is very important in determining the mechanical and chemical bonding at the interface of fiber to matrix. This is due to the larger surface area available on a rough fiber that can increase the adhesive bond strength by providing mechanical attaching locations on the fiber surface. A new type of eco-friendly brake friction composites containing flax fibers were developed. Optimization of the composition of friction composites reinforced by natural fibers met the requirements of automotive brake linings [76]. Experimental study on NFPC composites with polypropylene (PP) and polylactic acid (PLA) matrix, for friction properties indicates that natural fibers of vegetable and animal origin have a positive influence on final value of coefficient both for static and dynamic friction. Lowering of friction coefficient was achieved by adding the natural fibers – mainly flax fibers in NF compos-

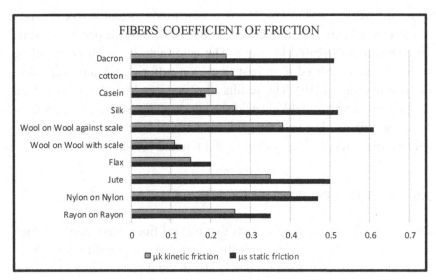

FIGURE 2.27 Fiber coefficient of static and kinetic friction.

ites with PP matrix, with increasing portion of natural fibers there was decreasing of friction coefficient – mainly static friction [77].

2.5 CHEMICAL PROPERTIES OF THE NATURAL FIBERS

Treatment of natural cellulosic fibers with NaOH effectively changes the surface morphology of the fibers and their crystallographic structure. The removal of surface impurities on plant fibers may be an advantage for fiber to matrix adhesion as it may facilitate both mechanical interlocking and the bonding reaction due to the exposure of the hydroxyl groups to chemicals, such as resins [78–79]. Hermetical alkali absorption process effectively removed the lignin and hemicelluloses from kenaf bast fibers at 160°C. The cellulose content of the fibers was 92%. The average surface hardness and elastic modulus of the fiber processed at 160°C yielded improvement of 348.1% and 111.3%, respectively, compared with those processed at 80°C. The increase of cellulose content of the processed fibers resulted in an improved fiber surface hardness and elastic modulus. The digestion temperature had a significant effect on tensile modulus and tensile strength properties of the fiber. When the digestion temperature increased from 110°C to 160°C, the tensile modulus and tensile strength of individual fibers were reduced by 42.8% and 22.9%, respectively, while the elongation increased by 1.1%. The acid hydrolyses fibers, cellulose whiskers, were isolated from bast fibers via sulfuric acid and hydrochloric acid hydrolysis [80]. The results illustrate that hydrolysis using both sulfuric acid and hydrochloric acid can successfully isolate whiskers from kenaf bast fibers. The sulfuric acid and hydrochloric acid whiskers have average diameters and length range of 3 nm and 100–500 nm.

2.6   THERMAL PROPERTIES OF THE NATURAL FIBERS

One of the distinctive properties is that natural fibers have lumen, which is hollow portion in fibers, indicating that green composites have pronounced thermal insulation property. The thermal conductivity of natural fiber composite is a function of the fiber volume fraction: it decreases with the increase of the fiber volume fraction due to the excellent thermal insulation properties [81].

KEYWORDS

- fatigue
- friction properties
- man made fiber physical properties
- mechanical properties
- natural fiber
- thermal properties

REFERENCES

1. Trade and Markets Division, Unlocking the commercial potential of natural fibers, market and policy analysis of non-basic food agricultural commodities team, FAO of the UN publications. [Online] 2012. http://www.fao.org/search/en/?cx=0181706201 43701104933%3Aqq82jsfba7w&q=BAST (accessed May 1, 2016).

2. Ashby, M. F.; Cebon, D. Materials selection in mechanical design, Troisieme Conference Europeenne sur les Materiaux et les Procedes Avances Euromat 93, Paris. [Online] June 8–10, 1993, Engineering Design Projects http://homepages.cae.wisc.edu/~me349/lecture_notes/material_selection.pdf (accessed May 6, 2016).

3. Material Selection Charts. [Online] http://www-materials.eng.cam.ac.uk/mpsite/interactive_charts/default.html (accessed May 6, 2016).

4. John, M. J.; Rajesh, D.; Anandjiwala, Recent Developments in Chemical Modification and Characterization of Natural Fiber-Reinforced Composites. Polym. Compos. February 2008, 29(2), 187–207.

5. *Agro Negocios*. [Online] Nov. 28, 2011 http://agronegocioscr.com/noticias/agronoticias/inta-y-mag-investigan-produccion-de-fibra-de-abaca (accessed May 6, 2016).

6. *Stipa tenacissima*. [Online] https://war.wikipedia.org/wiki/Stipa_tenacissima#/media/File:Stipa_tenacissima_1.JPG (accessed May 6, 2016).

7. Bagasse. [Online] https://commons.wikimedia.org/wiki/File:Sugarcane_bagasse.jpg (accessed May 6, 2016).

8. Bamboo garden. [Online] http://www.bamboogarden.com/care.htm (accessed May 6, 2016).

9. FAO Newsroom. [Online] http://www.fao.org/NEWSROOM/EN/news/2006/1000285/index.html (accessed May 6, 2016).

10. Natural fibers. [Online] http://www.fao.org/docrep/007/ad416e/ad416e06.htm (accessed May 6, 2016).

11. Crop Science. [Online] http://www.bayercropscience.cl/soluciones/fichacultivo.asp?id=129 (accessed May 6, 2016).

12. Hemp. [Online] https://en.wikipedia.org/wiki/Hemp (accessed May 6, 2016).

13. Agave. [Online] https://en.wikipedia.org/wiki/Agave_fourcroydes (accessed May 6, 2016).

14. FAO Newsroom. Concern at vanishing bananas. [Online] https://tenerifenaturewalks.files.wordpress.com/2013/03/p1140140.jpg (accessed May 6, 2016).

15. *Yucca faxoniana*. [Online] https://commons.wikimedia.org/wiki/File:Yucca_faxoniana_2.jpg (accessed May 6, 2016).

16. Jute. [Online] http://www.bangla-bagan.com/2015/08/13/growing-a-jute-plant-in-the-uk/ (accessed May 6, 2016).

17. *Ceiba pentandra*. [Online] https://en.wikipedia.org/wiki/Ceiba_pentandra (accessed May 6, 2016).

18. *Hibiscus cannabinus*. [Online] https://en.wikipedia.org/wiki/Kenaf (accessed May 6, 2016).

19. *Cannabis sativa*. [Online] https://commons.wikimedia.org/wiki/File:U.S._Government_Medical_Marijuana_crop._University_of_Mississippi._Oxford.jpg#/media/File:U. S._Government_Medical_Marijuana_crop._University_of_Mississippi._Oxford.jpg (accessed May 6, 2016).

20. *Urtica dioica*. [Online] https://upload.wikimedia.org/wikipedia/commons/a/aa/Stinging_nettle_plants.JPG (accessed May 6, 2016).

21. *Elaeis guineensis*. [Online] https://en.wikipedia.org/wiki/Palm_oil (accessed May 6, 2016).

22. *Attalea* (palm). [Online] https://commons.wikimedia.org/wiki/File:Attalea_funifera_Mart._ex_Spreng._(6709151365).jpg (accessed May 6, 2016).

23. Pineapple.   [Online]   https://commons.wikimedia.org/wiki/Category:Pineapple_
    fields_in_the_United_States#/media/File:Starr_020630–0021_Ananas_comosus.jpg
    (accessed May 6, 2016).
24. Phormium Amazing Red. [Online] https://commons.wikimedia.org/wiki/File:
    Phormium_Amazing_Red_1.jpg#/media/File:Phormium_Amazing_Red_1.jpg (ac-
    cessed May 6, 2016).
25. *Hibiscus sabdariffa.* [Online] https://commons.wikimedia.org/wiki/File:Roselle_(Hi-
    biscus_sabdariffa)_2.jpg#/media/File:Roselle_(Hibiscus_sabdariffa)_2.jpg (accessed
    May 6, 2016).
26. *Boehmeria nivea.* [Online] https://war.wikipedia.org/wiki/Boehmeria_nivea (accessed
    May 6, 2016).
27. Coir. [Online] https://en.wikipedia.org/wiki/Coir (accessed May 6, 2016).
28. Man-made fibers continue to grow, textile world. [Online] January/February 2015,
    http://www.textileworld.com/textile-world/fiber-world/2015/02/man-made-fibers-
    continue-to-grow/ (accessed May 6, 2016).
29. Statistics at FAO and the Statistical Program of Work. [Online] http://faostat3.fao.
    org/home/E (accessed May 6, 2015).
30. European Man-Made Fiber Association. [Online] http://www.cirfs.org/KeyStatistics/
    WorldManMadeFibersProduction.aspx (accessed May 6, 2016).
31. Bahia, S. Joint IGG on Hard Fibers and IGG on Jute, Kenaf and Allied Fibers, Joint
    Meeting of the 36th Session of the Intergovernmental Group on Hard Fibers and
    the 38th Session of the Intergovernmental Group on Jute, Kenaf and Allied Fibers.
    [Online] Nov. 16–18, 2011, http://www.fao.org/economic/est/est-commodities/
    jute-hard-fibers/jute-hard-fibers-meetings/igg37/igg36-fibers/en/ (accessed May 6,
    2016).
32. Asim, M.; Abdan, K.; Jawaid, M.; Nasir, M.; Dashtizadeh, Z.; Ishak, M. R.; Hoque,
    M. E. A Review on Pineapple Leaves Fiber and Its Composites, Int. J. Polym.
    Sci. [Online] 2015, Article ID 950567, 1–16. http://www.hindawi.com/journals/
    ijps/2015/950567/ (accessed May 6, 2016).
33. Yan, L.; Chouw, N. Sustainable concrete and structures with natural fiber reinforce-
    ments. [Online], Published by OMICS Group eBooks, http://www.esciencecentral.
    org/ebooks/infrastructure-corrosion-durability/sustainable-concrete-and-structures-
    with-natural-fiber-reinforcements.php (accessed May 6, 2016).
34. FOASTAT. [Online] http://faostat3.fao.org/download/P/PI/E (accessed May 6,
    2016).
35. Engelhart, A. W. The fiber year [Online], Issue 14, May 2014. http://clothesource.
    net/wp-content/uploads/2014/08/Fiber-Year-Review-summary.pdf (accessed May 6,
    2016)
36. Taj, S.; Munawar, M. A.; Khan, S. natural fiber-reinforced polymer composites, Proc.
    Pakistan Acad. Sci. 2007, 44(2), 129–144.
37. Al-Bahadly, E. A. O. The mechanical properties of natural fiber composites, PhD
    Theses, Faculty of Engineering Swinburne University of Technology, January, 2013.
38. Ku, H.; Wang, H.; Pattarachaiyakoop, N.; Trada, M. A review on the tensile proper-
    ties of natural fiber-reinforced polymer composites. Composites Part B 2011, 42,
    856–873.

39. Biagiotti, J.; Fiori, S.; Torre, L.; López-Manchado, M. A.; Kenny, J. M. Mechanical Properties of Polypropylene Matrix Composites Reinforced with Natural Fibers: A Statistical Approach. Polym. Compos. Feb., 2004, 26–36.
40. Westman, M. P.; Laddha, S. G.; Fifield, L. S.; Kafentzis, T. A.; Simmons, K. L. Natural Fiber Composites: A Review. National Technical Information Service, U.S. Department of Commerce, March 2010.
41. Zhu, J; Zhu, H.; Njuguna, J; Abhyankar, H. Recent Development of Flax Fibers and Their Reinforced Composites Based on Different Polymeric Matrices. Materials [Online] 2013, 6, 5171–5198. http://www.mdpi.com/journal/materials (accessed May 6, 2016).
42. Pereira, P. H. F.; Rosa, M. F.; Cioffi, M. O. H.; Benini, K. C. C. C.; Milanese, A. C.; Voorwald, H. J. C.; Mulinari, D. R., Vegetal fibers in polymeric composites: a review. Polímeros 2015, 25(1), 9–22.
43. Madsen, B.; Gamstedt, E. K. Wood versus Plant Fibers: Similarities and Differences in Composite Applications. Adv. Mater. Sci. Eng. [Online] 2013, 1–13 http://www.hindawi.com/journals/amse/2013/564346/ (accessed May 6, 2015).
44. Namvar, F.; Jawaid, M.; Tahir, P.; Mohamad, R. Potential use of plant fibers and their composites for biomedical applications. Bio-Resources 2014, 9, 5688–5706.
45. Schut, H. First Commercial Applications for Three New 'Eco' Fillers. Plast. Eng. [Online] July 30, 2010 https://plasticsengineeringblog.com/2010/07/30/first-commercial-applications-for-three-new-%E2%80%98eco%E2%80%99-fillers/ (accessed May 11, 2016).
46. Ververis, C.; Georghiou, K.; Christodoulakis, N.; Santas, P.; Santas, R. Fiber dimensions, lignin and cellulose content of various plant materials and their suitability for paper production. Ind. Crops Prod. 2004, 19(3), 245–254.
47. Huang, R.; Xu, X.; Lee, S.; Zhang, Y.; Kim, B.; Wu, Q. High Density Polyethylene Composites Reinforced with Hybrid Inorganic Fillers: Morphology, Mechanical and Thermal Expansion Performance. Materials [Online] 2013, 6, 4144–4138. http://www.mdpi.com/journal/materials (accessed May 6, 2016).
48. Faruk, O; Bledzki, A. K.; Fink, H.; Sain, M. Progress Report on Natural Fiber Reinforced Composites, Macromol. Mater. Eng. 2014, 299, 9–26.
49. Ishak M. R.; Leman, Z.; Sapuan, S. M.; Edeerozey, A. M. M.; Othman, I. S. Mechanical properties of kenaf bast and core fiber-reinforced unsaturated polyester composites. 9th National Symposium on Polymeric Materials (NSPM 2009), IOP Conf. Series: Materials Science and Engineering. [Online] 2010, 11, 1–6. http://iopscience.iop.org/1757–899X/11/1/012006/pdf/1757–899X_11_1_012006.pdf (accessed May 6, 2016).
50. Asim, M.; Abdan, K.; Jawaid, M.; Nasir, M.; Dashtizadeh, Z.; Ishak, M.; Hoque. M. A Review on Pineapple Leaves Fiber and Its Composite. Int. J. Polym. Sci. [Online] 2015, 1–16 http://www.hindawi.com/journals/ijps/2015/950567/cta/ (accessed May 6, 2016).
51. Davis, B. Natural Fiber-reinforced Concrete. [Online] 2007 http://people.ce.gatech.edu/~kk92/natfiber.pdf (accessed May 6, 2016).
52. Ramamoorthy, S. K.; Skrifvars, M.; Persson, R. A Review of Natural Fibers Used in Biocomposites: Plant, Animal and Regenerated Cellulose Fibers, Polymer Reviews.

[Online] 2005, 55(1), 107–162. http://www.tandfonline.com/doi/full/10.1080/15583 724.2014.971124 (accessed May 6, 2016).

53. Van Rijswijk, K.; Brouwer, B. D.; Beukers, A. Application of Natural Fiber Composites in the Development of Rural Societies. FAO Publications [Online] 2003. http:// www.fao.org/docrep/007/ad416e/ad416e00.htm (accessed May 6, 2016).

54. Monteiro, S. N.; Satyanarayana, K. G.; Ferreira, A. S.; Nascimento, D. C. O.; Lopes, F. P. D.; Silva, L. A.; Bevitori, A. B.; Inácio, W. P.; Bravo Neto, J.; Portela, T. G. Selection of high strength natural fibers. Matéria (Rio J.) [Online] 2010, 15(4). http:// www.scielo.br/scielo.php?script=sci_arttext&pid=S1517-70762010000400002 (accessed May 6, 2016).

55. Bosa, H. L.; Müssig, J.; Van den Oevera, M. Mechanical properties of short-flax-fiber reinforced compounds, Composites Part A: Applied Science and Manufacturing October, 2006, 37(10), 1591–1604.

56. Sparnins, E. Mechanical properties of flax fibers and their composites. PhD Theses, Division of Polymer Engineering Department of Materials and Manufacturing Engineering Lulea University of Technology S-971 87 Lulea, Sweden, 2006.

57. Diagram of the flax stem and fibers.svg. [Online] https://commons.wikimedia.org/ wiki/File:Diagram_of_the_flax_stem_and_fibers.svg (accessed May 6, 2016).

58. Sankar, P. H.; Reddy, Y. V. M.; Reddy, K H.; Kumar, M. A.; Ramesh, A. The Effect of Fiber Length on Tensile Properties of Polyester Resin Composites Reinforced by the Fibers of Sansevieria trifasciata. International Letters of Natural Sciences [Online] 2014, Vol. 8, 7–13 https://www.scipress.com/ILNS.8.7 (accessed May 6, 2016).

59. Mathew, L. Short-isora-fiber reinforced natural rubber composites, PhD Theses, Department of Polymer Science and Rubber Technology. Cochin University of Science and Technology, Kochi 22, Kerala, India 2006.

60. Singha, K.; Singha, M. Fiber Crimp Distribution in Nonwoven Structure. Frontiers in Science [Online] 2013, 3(1), 14–21 http://article.sapub.org/10.5923.j.fs.20130301.03. html (accessed May 6, 2016).

61. Piyatuchsananon, T.; Furuya, A.; Ren, B.; Goda, K. Effect of Fiber Waviness on Tensile Strength of a Flax-Sliver-Reinforced Composite Material. Adv. Mater. Sci. Eng. [Online] 2015, 1–8 http://www.hindawi.com/journals/amse/2015/345398/ (accessed May 6, 2016).

62. Bauer-Kurz, I. Fiber crimp and crimp stability in nonwoven fabric processes. PhD Theses, NCSU, USA, 2000.

63. El Meessiry, M.; Abd-Ellatif, S. Characterization of Egyptian cotton fibers, IJFTR March 2013, Vol. 38, 109–113.

64. Pickering, K. L.; Efendy, M. G.; Le, T. M.; A review of recent developments in natural fiber composites and their mechanical performance. Composites Part A [Online] 2016, 83, 98–112 http://www.sciencedirect.com/science/article/pii/ S1359835X15003115 (accessed May 6, 2016).

65. Nuthonga, W.; Uawongsuwanb. P.; Pivsa-Art, W.; Hamadab, H. Impact property of flexible epoxy treated natural fiber-reinforced PLA composites. Energy Procedia [Online] 2013, 34, 839–847. http://www.sciencedirect.com/science/article/pii/ S1876610213010631 (accessed May 6, 2016).

66. Rowell, R.; Han, J.; Rowell, J. Characterization and Factors Effecting Fiber Properties, Natural Polymers and Agro fibers Composites. Sãn Carlos – Brazil – 2000,

115–134. Editors: Frollini, E.; Leão, A. L.; Carlos, L. H. C. M.: USP-IQSC/Embrapa Instrumentação Agropecuária/Botucatu: UNESP, 2000.

67. Xu, Y. Creep behavior of natural fiber-reinforced polymer composites. PhD Theses, The School of Renewable Natural Resources, Louisiana State University, USA, 2009.

68. Belaadia, A.; Bezazia, A.; Ourchak, B.; Scarpac, F. Tensile static and fatigue behavior of sisal fibers. Mater. Des. April 2013, 46, 76–83.

69. Shah, D. U.; Schubel, P. J.; Clifford, M. J.; Licence, P. Fatigue life evaluation of aligned plant fiber composites through S-N curves and constant-life diagrams. Compos. Sci. Technol. 2013, 74, 139–149.

70. Vasconcellos, D.; Touchard, F.; Chocinski-Arnault, L. Cyclic fatigue behavior of woven hemp/epoxy composite: damage analysis. 15th European conference on composite materials, Venice, Italy, 2012.

71. Towo, A. N.; Ansell, M. P.; Fatigue evaluation and dynamic mechanical thermal analysis of sisal fiber–thermosetting resin composites. Compos. Sci. Technol. 2009, 68(3–4), 925–940.

72. Blaadi, A.; Bezazi, A.; Bourchak, M.; Fabrizio, S. Tensile static and fatigue behavior of sisal fibers, Mater. Des. April 2013, 46, 76–83.

73. Shahada, A. Isaac, D. H. Fatigue Properties of Hemp and Glass Fiber Composites. Polym. Compos. 2014, Vol 35, issue 10, 1926–1934.

74. Liang, S.; Gning, L.; Guillaumat, L. Fatigue behavior of flax/epoxy composite, Compos. Sci. Technol. 2012, 72, 535–543.

75. Karman, M. Fiber Friction. [Online], http://uctemtn.blogspot.com.e.g./p/fiber-friction.html (accessed May 6, 2016).

76. Fu, S.; Suo, B.; Yun, L.; Lu, Y.; Wang, H.; Qi, S.; Jiang, S.; Lu, Y.; Matejka, V. Development of eco-friendly brake friction composites containing flax fibers, J. Reinf. Plast. Compos. April 2012, 31(10), 681–689.

77. Běhálek, L.; Lenfeld, P.; Said, M.;Bobek, J.; Ausperger, A. Friction properties of composites with natural fibers, synthetic and biodegradable polymer matrix. NANCON Conf., Brno, Czech Republic, EU, 2012.

78. Mwaikambo, Y.; Ansell, M. P. The effect of chemical treatment on the properties of hemp, sisal, jute and kapok fibers for composite reinforcement. 2nd International Wood and Natural Fiber Composites Symposium, Kassel/Germany, 1999.

79. Shi, L. J.; Shi, S. Q.; Barnes, H. M.; Horstemeyer, M.; Wang, J.; Hassan, E. L. Kenaf Bast Fibers—Part I: Hermetical Alkali Digestion. Int. J. Polymer. Sci. [Online] 2011, 1–8 http://www.hindawi.com/journals/ijps/2011/212047/ (accessed May 15, 2015).

80. Zaini, L.; Jonoobi, M.; Tahir, P.; Karimi, S. Isolation and Characterization of Cellulose Whiskers from Kenaf (*Hibiscus cannabinus* L.) Bast Fibers. Journal of Biomaterials and Nano biotechnology [Online] 2013, 4, 37–44. http://www.scirp.org/journal/PaperInformation.aspx?PaperID=26847 (accessed May 20, 2015).

81. Osugi, R.; Takagi, H.; Liu, K.; Gennai, Y. Thermal Conductivity Behavior of Natural Fiber-Reinforced Composites. Asian Pacific Conference for Materials and Mechanics, Yokohama, Japan, 2009. http://www.jsme.or.jp/conference/mmdconf09/data/abst/a163.pdf (accessed May 6, 2016).

PART II

DESIGN OF NATURAL FIBER POLYMER COMPOSITES

CHAPTER 3

NATURAL FIBER REINFORCEMENT DESIGN

CONTENTS

3.1 INTRODUCTION

The objective of forming textile composite is to use the textile fibers as reinforcement of composite matrix in order to get: high strength, light weight, consolidation part with net shape to fit the designed part. The textile composite is formed from two or more materials in different proportions that gives the final composite the designed properties, such as mechanical properties, thermal expansion, etc. The natural fibers used for composite applications are mainly: Bast fibers, such as flax, hemp, kenaf, ramie, jute, nettle, etc., Leaf fibers, like sisal, henequen, pineapple, manila, Fruit fibers, like coir, Grass, such as bamboo, wheat, rice, oat, barley, elephant grass, and others are stalks of the different plantations, as illustrated in Figure 3.1.

A composite structure consists of three components: flax fibers, flax woven fabric and wheat straw as reinforcement and using a natural resin as a matrix, for instance as shown in Figure 3.2 an example of biodegradable composite [1].

Wood-polymer composite (WPC) industry have boomed in the last decade. Applications for these composites include a variety of building, consumer, industrial and automotive products. Polymer composites made with wood and natural fibers, such as rice hulls, flax, hemp, jute, or kenaf are environmentally sound due to their biodegradability [2]. There are today huge unused quantities of agricultural (straw) residues around the globe. In 2014, the annual global production of wheat was around 705 million tons, out of which 75% was produced by 18 countries, while around

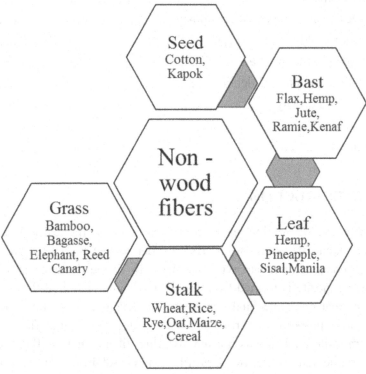

FIGURE 3.1   Non-Wood fibers for composite applications.

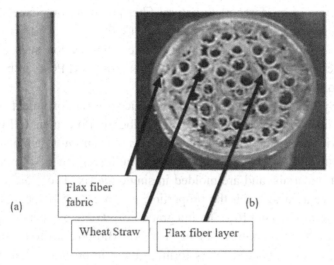

(a)

Flax fiber fabric

(b)

Wheat Straw

Flax fiber layer

FIGURE 3.2 Natural fiber biodegradable composite.

20% was produced by EU [3]. According to the statistical data, the average yield of straw is around 1.3–1.4 kg per kg of grain [4], every kilogram of grain harvested is accompanied by production of 1–1.5 kg of the straw. The wheat and rice straw statistics indicates that there are about 710 million metric tons of wheat straw and 650–975 million ton of rice straw each year [5].

3.2 CLASSIFICATION OF COMPOSITE

The structure of the composite consists of three dimensional preforms in the form of the designed part emerged compositely by matrix. The fiber reinforcement can be in arrangement of fibrous structure of individual fibers, yarns, weave structure, woven, knitted, braided, nonwoven and tri-axial fabrics.

The matrix holding the fibrous structure in pre-determined form, transfers the applied loads in all the directions to all the fibers in the preform, and protects the reinforcement from environment effect.

The composites are classically classified according to:
• Type of matrix: polymer matrix composites (PMC), metal matrix composites (MMC), and ceramic matrix composites (CMC).

- Type of reinforcement: natural fiber polymer composite (NFPC), and wood fiber polymer composite (WPC).

The use of cement matrix adds another abbreviation such as wood cement composites (WCC). The textile composites (FPC) can be classified according to the type of fibers used as:

- **Synthetic fiber composite**, synthetic fiber reinforcement (SFOC);
- **Natural fiber composite**, natural fiber reinforcement (NFPC).

Hybrid composites, which may be formed from multi laminates reinforcement built of different material or reinforcement of a mixture of different materials, and are molded in single matrix. These technologies usually are used to reach the properties, which cannot be obtained with a single reinforcement [6]. The integration of variety of fibers in a single matrix results in the development of hybrid Bio composite by blending different types of natural fibers forming the reinforcements, or forming layers of fibers in form of mat or fabric, or forming an nonwoven fabric of different blend of fibers. In some designs of hybrid composites, natural fibers with manmade fibers are used. For instance, hybrid composites are Palmyra/glass, bamboo/glass, jute/glass, jute/biomass, sisal/kapok and kenaf/glass.

The natural fiber polymer composites may be biodegradable (bio-composite) or partially biodegradable, depending on the matrix material:

- Natural base polymer;
- Biodegradable petroleum base;
- Partial biodegradable;
- Non-biodegradable petroleum base.

The textile composite also may be classified according to the fiber dimensions and orientation of the fibers in the reinforcement, as demonstrated in Figure 3.3, into:

- Particles;
- Fibers;
- Yarns;
- Fabrics.

Classification according to the type of reinforcement:

- Discreet (particles, whiskers, short fibers);
- Preform construction (short fiber, continuous yarns, 2-D fabrics, 3-D fabrics);

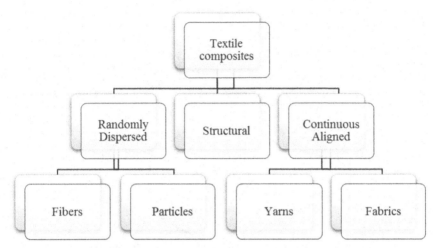

FIGURE 3.3 Classification of textile composites.

- Fiber orientation (uncontrolled random, one direction, orthogonal, 3-D orientation);
- Degree of fiber entanglements (non, twisted, crossed).

The final properties of the composite will depend on the structural alignment of the filler in the space of the composite. The criteria for the choice of suitable reinforcing fiber properties include: tensile and stiffness, elongation at failure, thermal stability, adhesion of the fibers and matrix, dynamic behavior, long-term behavior, and degradability. The reinforcement is responsible to give the architectures of the composite designed part shape. In most cases, its material is different from that of the matrix, making composites strongly heterogeneous materials. Nano-particles are presently considered as a high-potential filler materials for the improvement of the mechanical and physical properties of polymer composites [7].

Figure 3.4 illustrates the classification of the textile composite preforms, accordingly to the type of material used as reinforcement.

The composite can be formed with the continuous filament yarns, aligning in the direction of the maximum load in one or multi-layer, as shown in Figure 3.5a. The fibers may be cut into predetermined length and blended with the matrix material to form the final form of the composite, Figure 3.5b. WPC use wood-flour or soya beans as filler in thermoplastic matrix, Figure 3.5c, with size from 50 to 300 micron.

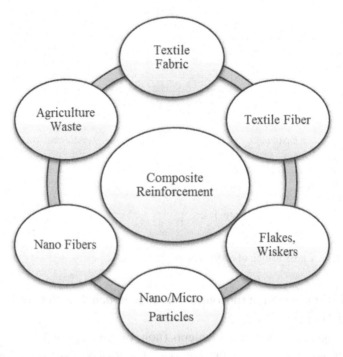

FIGURE 3.4   Classification of textile composite reinforcement.

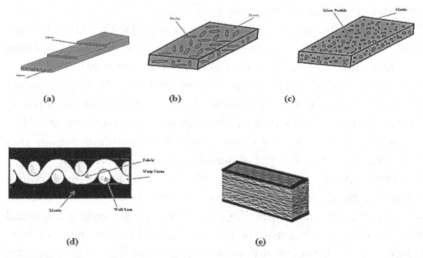

FIGURE 3.5   Composite construction.

The wood-polymer composite impact strength depends on the wood flour particle size and the wood flour content percentage, especially for course mesh of wood-flour. Thus, the composite properties may vary in every point so that it can be considered as anisotropic. Fabric can be used as the reinforcement of the composite in one layer, Figure 3.5d, or multi-laminate, Figure 3.5e.

3.3   DESIGN TECHNIQUES OF COMPOSITE

The natural fibers are generally anisotropic material in all properties, tensile, impact, specific gravity, thermal, etc. The stress strain behavior of the composite component represents a problem: how to interpret for this variation of mechanical properties of the reinforcement and matrix [8], designed for structural performance and particularly, the mechanism in damage and failure for composite. The determination principles and methods for natural fiber composites are different from the metal materials [9]. The design allowable values need to use design stress analysis and determine the geometry details. For material choice the following allowable values should be determined: the related static strength, fatigue strength, damage tolerance, according to the end use of the composite. Also, the manufacturing technique of the part affects the final mechanical properties due to the distribution of fibers with different configuration of fiber orientation during molding. When using thermoplastic matrix, the shrinkage of the composite will be very significant because of the fiber orientation variability. The variation of fiber orientation will make the polymer shrinkage differ in X, Y, Z directions due to the resistance to the polymer flow. NASA [10] gives the following requirements for the designer to define: environmental and performance requirements, choice of materials, the intended manufacturing techniques, test geometric configurations, and define nondestructive evaluation (NDT). Finally, the designer should analyze, weight, and balance the above factors to produce an initial design concept as illustrated in Figure 3.6.

Each of the requirements of final composite is a function of the both – the fiber and matrix properties and technique of its manufacturing. Table 3.1 gives the general factors which reflect the effect of material properties on the product specifications.

FIGURE 3.6   Main analysis of material choice.

TABLE 3.1   Material Selection for Composite Design

Composite requirements	Fiber properties			
Performance	Fiber strength & elongation	Fiber stiffness	Fiber impact strength	Fiber and matrix creep properties
Weight	Fiber density	Size		
Recycling	CO_2 emission	Possibility of reuse	Safe disposal	
Bio-degradation	Environmental effect on fiber and matrix degradation	Rate of degradation		
Environmental issue	Energy consumption in composite elements manufacturing and forming	CO_2 emission		
Cost	Raw material cost	Manufacturing cost		

3.4   CHOICE OF THE FABRIC TYPE

Continuous fiber reinforced composites are now firmly established engineering materials for the manufacture of components in the automotive and aerospace industries [8]. In this respect, composite fabrics provide flexibility in the design manufacture. Many investigations indicate that textile fabrics possess hardly any stiffness in bending or compression, so it is able to cover a 3-D body gracefully and can deform to a complex shape easily. However, because of its small bending rigidity, it cannot support compressive stresses. When sheet of textile material deforms in a compressive direction, buckling occurs and wrinkles are formed. Wrinkling (i.e., buckling of fibers), arising during the forming of textile composites, tends to significantly degrade the performance characteristics of the final product produced net shape reducing overall manufacturing cost, but also in the versatility of the weaving process [11]. That 3-D-fabric reinforcement can overcome the problems of low inter-laminar and through-thickness strength typical for traditional 2-D-laminates has been known for many years. The mechanical properties of composites made from fibers laminate are characterized by high in-plane stiffness and strength and lower out-of-plane stiffness and strength. [12]. One of the main advantages of 3-D-fabric reinforcements is the increased fracture toughness. The fracture toughness is a measure of how well a material containing cracks can resist fracture. The increased resistance to delamination also has a positive effect on compression strength after impact [13]. The deformation modes of composites fabrics during the forming process are different than those of sheet metal, a number of deformation mechanisms are available, including shear deformation between warp and weft fibers, fiber straightening, relative fibers slip and yarn buckling [8]. The use of geometry of deep-drawing tools the shear angle may reach 38°, and the contraction in the fabric due to yarn stress will lead to the distribution of strain to vary in the area surrounding the 3-D shapes [4, 14] as shown in Figure 3.7.

Consequently, any type of fabric can be used with the suitable method when it possesses the following properties [8]:

- Low fabric bending resistance;
- Anisotropy in mechanical properties;
- Low between yarn coefficient of friction;

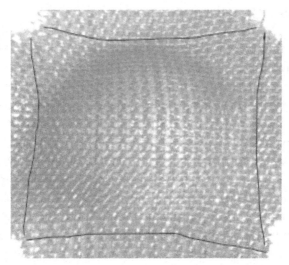

FIGURE 3.7   Fabric deformation during forming.

- Low shear modulus;
- High value of bursting strength;
- High value of melting temperature.

For knitted fabric, shear modulus is much lower than the woven or braided fabric, due to its loose structure and the difference of the mechanical properties in the direction of walls and courses. This makes it more suitable for the use in 3-D preform formations. Moreover, the use of Lycra yarn in either half plated of full plated knitted fabric gives more decrease of the shear modulus. Thus, the most suitable type of fabric for the manufacturing of 3-D complicated shapes for composite application are the knitted [14]. Some types of composite are designed using multi laminates to reach their proposed requirements as illustrated in Figure 3.5e. The composite is made from laminates of several layers of woven fabric. The direction of the warps in each laminate may be unidirectional or will take different angles, as shown in Figure 3.8.

The delamination failure of textile composite will be more noticeable in hybrid composites, where several reinforcements of different materials are presented in single matrix or when one reinforcement is used with a mixture of different materials [15].

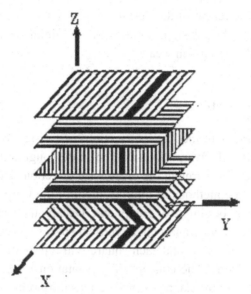

FIGURE 3.8 Laminate architect of the composite.

3.5 DESIGN OF LAMINATES

The yarns or tow are used for formation of non-interlaced 2-D laminate. Fabrics are manufactured by the weaving, knitting, triaxial, warp knitting, braiding technology. Non-woven structures are also utilized for laminate formation.

3.5.1 ONE DIMENSIONAL

Continuous filaments, yarns or tow are oriented in one direction and laid in X-Y plane, forming the reinforcement structure of a composite.

3.5.2 2-D FABRIC

In order to express the fabric design, the weft yarns are presented by horizontal lines and the warp yarn by vertical lines. The intersections of these lines will represent the crossing of the yarns over each other or the yarn intersections. The repeat of the design is usually the unit of intersections representing the fabric design, it indicates the minimum number of warp and weft threads for a given weave.

3.5.2.1 Basic Weave

The woven fabric is formed by intersections of two sets of perpendicular yarns, warp and weft. Warp yarns are run in the longitudinal direction of the fabric while weft yarns in the cross wise direction. The woven structures which are the methods of interlacement of the warp and weft yarns in one repeat, may be simple or compound. Usually the weave structure is presented by the square paper where vertical spaces represent warp yarns and horizontal weft yarns and each square will represent intersections of the two sets of yarns. The concept of presentation is each blank square represents weft yarn overlapped warp yarn, as illustrated in Figure 3.9.

The design repeat is the minimum number of warp and weft yarn interlacements that repeated in the fabric design.

3.5.2.2 Knitting Fabric

Knitting process is completely different from weaving process, fabric is formed by interlacing of looped yarns as illustrated in Figure 3.10. In the

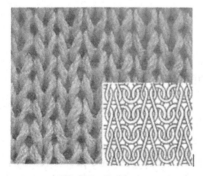
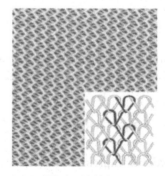

Weft knitting   **Warp knitting**

FIGURE 3.9   Woven fabric.

case of warp knitting, loops are formed in vertical direction. Single yarn looped horizontally forms a raw of loops (course), as the knitting proceeds, vertical loops forming the walls. In warp knitted fabrics the loops are formed vertically. Different structures of the knitted fabrics are illustrated in Figure 3.11.

3.5.2.3 Triaxial Fabric

2-D fabric consists of two sets of yarns (orthogonal sets of yarns) intersected at 90 degrees on each other, biaxial woven fabric. These yarns may be interlaced to form integral structure of thickness function of the weft

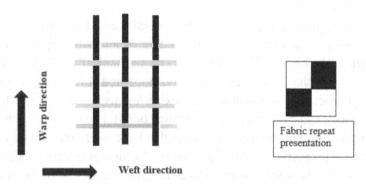

FIGURE 3.10   Basic knitted fabrics. (a) Warp knitting; (b) Weft knitting.

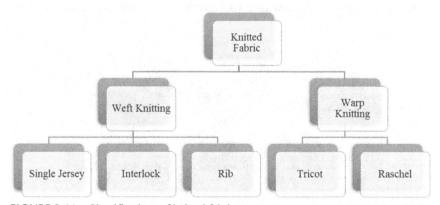

FIGURE 3.11   Classifications of knitted fabrics.

and warp yarns diameters as shown in Figure 3.9. In the case of triaxial fabric, three sets of yarns run in three directions to form equilibrium triangle, as shown in Figure 3.12. The number of yarns per cm in each direction gives several shapes of final fabric. The advantages of the triaxial fabric are: light areal weight, low material cost, isotropic properties, high shear resistance, and bursting strength. Such fabric has a lower crimp than woven fabric. Several structures are produced, as illustrated in Figure 3.12. It is shown that triaxial woven fabrics exhibit a relatively higher and more uniform resistance to extension, shear deformation, and burst deformation than comparable biaxial woven fabrics [16, 17].

3.5.2.4 Nonwoven Fabric

Nonwoven fabric can be processed by several methods for binding a mat of randomly oriented fibers together to give its integrity and mechanical properties. The binding of the fibers may be mechanically (Needle punch), chemically, or thermally. There are several methods of forming the nonwoven fabric: dry laid nonwovens, extrusion technology (spun bond, melt blown), and hybrid technology (Stitch bond). Each method ended by completely different material structures that have a different physical and mechanical properties. The production rate the highest is spun bonded and most popular for needle punched. Needle punching is one of the widely used techniques for the production of the nonwoven fabrics, which consist of web forming, needle bonding. The hybrid technology base includes combining systems at least one basic nonwoven web formation to join

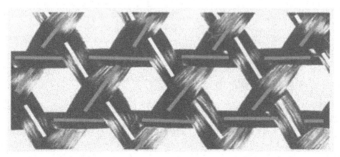

FIGURE 3.12   Triaxial fabric.

two or more fabric. Nonwoven methods are the least versatile, but fastest of the other manufacturing technologies. Figure 3.13 illustrates the basic principle of needle punching method.

Regarding natural fibers, nonwoven technology can be used to manufacture the fiber reinforcement laminates, even in the case of bast fibers, which can be manufactured by diverse techniques, such as stitch-bonding, hot calendaring, needle punching, hot-air thermal bonding, oven bonding, hydro entanglement, etc. Needle punching nonwoven fabrics from bast fibers were used for the production of reinforcement of composite preform. Jute needle punched nonwoven fabric gives better results than the jute woven fabric, due to better fiber-resin interface strength [13, 16–20]. The nonwoven preforms can be produced with different thickness, as required by the application, with no need to be made from several laminates as for woven fabric. The mechanical properties such as strength, elongation, bursting strength, bending rigidity and porosity of the jute nonwoven will depend on the weight per unit area of the nonwoven fabric and the machine settings. Needle-punched nonwoven may be a successful reinforcing agent for the jute-based composites [20] (Figure 3.14).

3.5.2.5 Braiding Fabric

Braiding as one of the key preforming technologies for producing low-cost high volume composites in industries such as automotive [21]. Braided composites have superior toughness and fatigue strength in comparison to

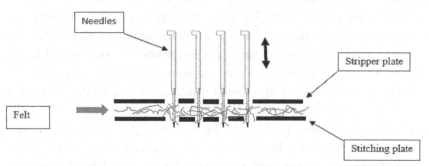

FIGURE 3.13   Basic principle of nonwoven needle punching fabric formation.

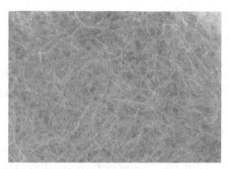

FIGURE 3.14   Needle punched jute nonwoven mat.

filament wound composites [22]. Braids can be produced either as seam-less tubes or flat fabrics.

The technique of braiding is very old and has been used by the ancient civilizations back to 4000 BC in manufacturing of hair forming, ropes and fabrics. The basic idea of braided structures is when three or more yarns intertwined in such way that no two yarns are twisted around another. Braiding processes have the ability to fabricate a wide range of complex shapes. Braiding process can form shapes either by over-braiding man-drels on conventional machine (two-dimensional process) or by using new braiding processes to form solid shapes (three-dimensional process). 3-D braided fabrics have found the applications in the areas including medi-cine, aerospace, automobiles, train components, and reinforced hoses [23]. The initial development of 3-D braided fabrics came from the composite and medical industries. 3-D braided fabrics can be manufactured in myr-iad varieties of cross-sections, and their near-net complex shapes made it possible to design very specialized products for both industries.

The braided fabric is flexible before formation, and thus the fabric can conform to various shapes. The braided fabric may be formed around a mandrel, and rather complex shapes can be formed. The principle of braid-ing is illustrated in Figure 3.15, the minimum numbers of yarns are 3.

If the yarn interlacing in weaving is orthogonal, in braiding the angle between the yarns is less than 90°. The interlacing sequence in flat or cir-cular braid is the same. Figure 3.16 illustrates samples of braided weave.

The structure of the braided fabric may be classified similar to the woven fabrics into: plane, twill 1/2 or twill 1/3 to explain the intersections of the yarns. Figure 3.17 illustrates the interlacement order in different designs.

The structure of the braid can be modified through the modification of the movement of the yarns when interlaced with the other yarns resulting in several designs, as illustrated in Figure 3.18.

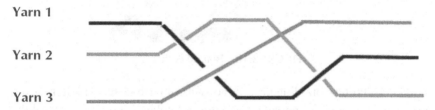

FIGURE 3.15   Path of the interlaced yarns in braided structure.

FIGURE 3.16   Braided weave.

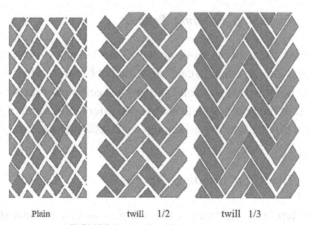

Plain   twill   1/2   twill   1/3

FIGURE 3.17   Braided structures.

FIGURE 3.18   Braided designs.

The 3-D rotary braiding process consists of base plates with horn gears and mobile bobbins arranged upon them. Switches are used to control the position of the threads and horn gears. In circular braiding, the yarn bobbins (with opposite directions of rotation) will form the finale yarn reinforcement shape on the surface of mandrel, determine the yarn's pattern and crossing point. At this crossing point, the bobbins change their path to produce the upper and inner side of the braid.

Each design of braid will give different fabric physical and mechanical properties. Generally, the circular braiding process produces braids with rotational symmetry.

Another principle of the motion of the bobbins of the yarns forming the braid are:

3.5.2.5.1 Four Step Braiding Process

In this process, the bobbins move on the X and Y axes, which are mutually perpendicular to each other. In each step, the bobbins move to the neighboring crossing point in two ways and stop for a specific interval of time. Basic arrangement of the braiding field is obtained after a minimum of four steps. This method produces braids which have a constant cross section.

3.5.2.5.2 Two Step Braiding Process

In the two-step braiding process, the bobbins move continuously without stopping. They move on the track plate through the complete structure and around the standing ends, such that the movements of bobbins are faster when compared to the four-step braiding process. The bobbins can move only in two directions, so the process is called the two-step braiding process.

3.5.2.5.3 Tubular Braided Fabric

Tubular 3-D braided preforms, is processed on using two or four step process to braid the yarns on surface of a mandrel with the required final shape. Figure 3.19 illustrates a tubular braid [23, 24]. Braided structures tend to allow high flow of resin during molding processing and can conform to the mold well [25]. The tubular braided fabric can be flattened or circular formed around a mandrill with the required shape more over an axial yarn can be used to give final form strength and stiffness.

FIGURE 3.19   Tubular braiding.

3.5.3 3-D FABRIC

The problem of having a heavy weight preforms (1500 g/m^2), which the high weight of the fabrics, together with the ability to orientate the fiber at different angles, enables fewer layers to be used is fully solved through the use of 3-D fabric. 2-D fabric can be laid in several layers to reach the required areal density, however this solution is of limited application due to the delamination problem and the presence of voids. In recent years 3-D fabrics have begun to find favor in the construction of composite components. The target objectives of the 3-D preform is to manufacture a preform with the final neat yarn 3-D space structure ready for the manufacturing the composite part. 3-D fabrics are attractive to the composite industry because of the possibility of obtaining near-net shaped and delamination resistant reinforcements directly. Moreover, the 3-D preform technology facilitates to use automation and CAD/CAM systems for the preform manufacturing. While preforms produced from traditional 2-D laminates can only be processed into relatively simple and slightly curved shapes, preform for a composite component with a complicated shape can be made to the near-net-shape with 3-D weaving. This ability of 3-D weaving producing near-net-shape preforms can reduce the production costs thanks to the reduction in material wastage, need for machining and joining, and the amount of material handled during lay-up [26].

3.5.3.1 Classification of the Different 3-D Preforms

One of the main advantages of 3-D-fabric reinforcements is the increased fracture toughness. The fracture toughness is a measure of how well a material containing cracks can resist fracture. The increased resistance to delamination also has a positive effect on compression strength after impact [13, 26, 27]. Manufacturing methods of 3-D preforms can be classified as, Braiding, Weft knitting, Warp knitting, Weaving and stitched assemblies. The basic principles to produce thick mat of fibrous structure are given in Table 3.2. The 3-D textile materials offer particular properties along all three-dimensional axes that are not achievable with 2-D laminate and other reinforcements. This integrated architecture provides

TABLE 3.2   Classification of 3-D Fabric for Composite Reinforcement [13, 26, 27]

Process	Principles	Remarks
1. Multiaxial knitting fabrics	• Weft knitting fabric with axial warp yarns	• Less flexibility • Limited hollow preforms • Limited strength in z-direction

2. 3-D-contoured weaving	• Weaving machine:
	  - dobby loom
	  - jacquard loom
	  - 3-D Weaving loom
		• Limited shapes
		• Low productivity
		• Profile with constant cross section

3. Circular tubular woven fabric	• Circular loom	• Limited shapes • High production rate
4. Integral fabric	• Multilayer fabric • 3-D orthogonal fabric • 3-D woven fabric • Multi-axial fabric	• Less flexibility • Limited preform shapes
5. Distance fabric	• Distance fabric with interlocking yarns • Distance fabric with plane interlock	• Less flexibility • Limited preform shapes
6. Multi-layer nonwoven knitted interlacing	• Weave & Stitch • Simultaneous stitch	• Low stiffness • Less flexibility • Limited preform shapes
7. Nonwoven	• Needle punching • Stitch bonding • Hydroentanglment • Adhesive bonding • Thermal bonding • Spun-laid	• High productivity • Non crimp fiber assembly • low mechanical properties • Low shear modulus
8. 3-D Knitted fabric	• Weft knitting flatbed (thick reinforcement, spacer fabric) • Circular tubular spacer tubular fabric	• Warp Knitting: limited applications • Pipe Y or T shapes • High production rate

TABLE 3.2   (Continued).

Process	Principles	Remarks
9. Braiding	• Flat braided • 3-D Braided shapes	• Less flexibility • Limited shapes • Low shear modulus
10. Stitched assembly	• Fiber assembly stitched bonded • Assembly of yarns in multiple layers • Assembly of fabric in multiple layers	• Limited flexibility • Limited strength in z-direction • High heterogeneity in properties

improved stiffness and strength in the transverse direction and hinders the separation of in-plane layers in comparison to traditional multi-layer 2-D fabrics. Because of their high transverse strength, high shear stiffness, low delamination tendency 3-D textile composites from weaving and knitting have received attention recently [28].

Figure 3.20 illustrates the ranking of the three designs of laminate pre-forms, using fiber tows, 2-D weave and 3-D weave, taking into consideration preform properties and manufacturing parameters.

3.5.3.2   Some Methods of Manufacturing the 3-D Fabric

a. Woven 3-D Fabrics

- The conventional 2-D weaving machine uses three sets of yarns (ground warp, pile warp and pile weft) to produce pile fabrics.
- The conventional 2-D weaving machine is designed to interlace two orthogonal sets of yarns (warp and weft) with an additional set of yarns functioning as binder warps or interlace yarns in the through-the-thickness or Z direction, either layer to layer or through the layers, Figure 3.21a,b.

b. Non-Interlaced Multiaxial Orthogonal 3-D Fabrics

Special weaving machines were designed to produce multiaxial orthogonal 3-D fabrics, as illustrated in Figure 3.21c. The main feature

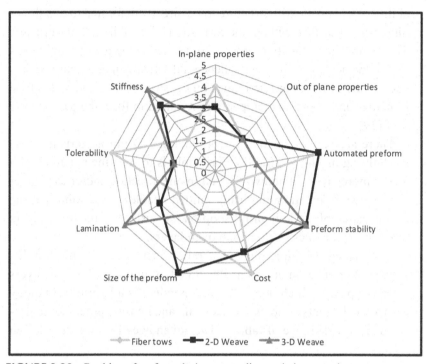

FIGURE 3.20   Ranking of preform designs according to their properties.

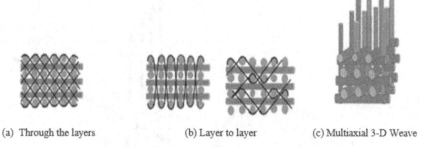

(a) Through the layers   (b) Layer to layer   (c) Multiaxial 3-D Weave

FIGURE 3.21   Principles of 3-D woven fabric.

of these structures is the yarns are non-interlaced, which effects the mechanical properties of the final structure.

c. Knitted 3-D Fabrics

Knitted machine have received much attention in recent years in the composite field due to their excellent formability. Formability is the ability of a planar textile structure to be directly deformed to fit a three-dimensional surface without the formation of wrinkling, kinks or tears. The composite production process to be used is the preliminary attention in the design of composite, hence individually manufacturing process imposes particular restrictions on the structural design on the type of matrix and fibers, the temperature required to form the part and the cost [29].

There are various method for producing 3-D fibrous structures on the knitting machine. The basic principle is to feed the machine with one or more layers of nonwoven mat in one direction and/or each layer has its own orientation and using knitted needles to pass through the fibrous assembly to knit yarn's loop on the surface of the outer layers through the thickness of the nonwoven layers.

- Multiaxial: Multiple plies of parallel fibers, each laying in a different orientation or axis as illustrated in Figure 3.22. These layers are typically stitch bonded. Stitch bonding is a hybrid technology using elements of nonwoven, sewing and knitting processes to produce a wide range of fabrics. The advantage of such fabric design

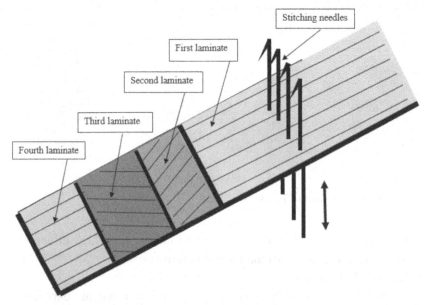

FIGURE 3.22   Multiaxial fabric.

that there is no crimped fibers to assist the resin flow during forming of the composite, besides the stitch bonding will increase the speed of resin fusion.

- Stitch bonding: Stitch bonding is feeding a cross laid web of fibers or nonwoven mat into warp knitting. Stitch yarn will move through the web from one side to the other binding the fibrous web. Several principles of binding technique are used such as:
  - Stitch bonding of webs with stitching threads, Figure 3.23a;
  - Stitch bonding of webs without stitching threads using the fibers presented in the web itself;
  - Stitch bonding of webs with loop threads, Figure 3.23b.
- Spacer fabrics: the spacer fabric is a three dimension fabric formed from two surface fabrics separated by a pile yarns, keeping a space as illustrated in Figure 3.24. There are several methods of manufacturing the spacer fabric, such as: weaving, braiding, warp knitting, weft knitting.

Flat knitted spacer fabrics offer a strong potential for complex shape preforms, which could be used to manufacture composites with reduced waste and shorter production times. A reinforced spacer fabric made of individual surface layers and joined with connecting layers shows improved mechanical properties for lightweight applications, such as textile-based sandwich preforms [30, 31]. The advantage of knitted spacer fabric was found to be high preform performance, flexibility in design, cost efficient.

d. Triaxial Braided

Triaxial structure is formed when the additional axial yarns are introduced into a 2/2 repeating pattern of the braided fabric as illustrated in Figure 3.25. This can be manufacture by two or four step braiding process.

3.6 FABRIC STRUCTURE CALCULATIONS

The main basic formulas for the calculation of the designed fabric for the composite fiber reinforcements are presented in the following subsections.

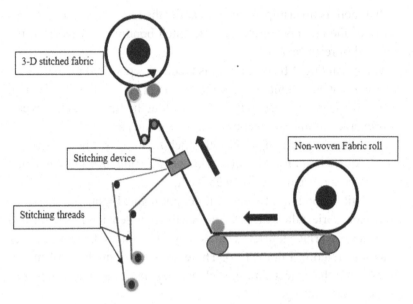

(a) Stitch bonding

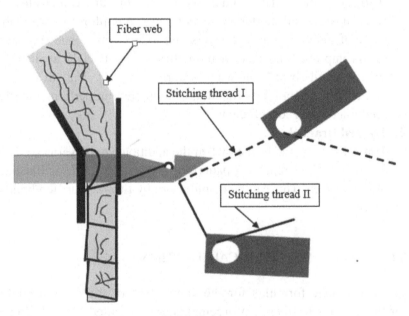

(b) Maliwatt

FIGURE 3.23   Stitch bonded fabric. (a) Stitch bonding; (b) Maliwatt.

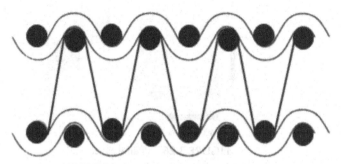

FIGURE 3.24   Principles of spacer fabric.

FIGURE 3.25   Triaxial braided structure.

3.6.1   WOVEN FABRIC DESIGNS

The presentation of the plain weave is given in Figure 3.26. There are three types of weaves that are known as basic weaves, which include *plain weave* (the simplest and smallest repeat size possible) and its derivatives, *twill weaves* and their derivatives, and *satin weaves* and their derivatives. However, they form the base for creating any complex/intricate structures (such as multi-layer fabrics and pile weave structures) and weaves with extremely large patterns that are known as Jacquard designs.

The weave design can be written as 1/1 for plain weave or 1/3 for twill or 1/5 satin weave. Basket weave is a derivative of plain weave 2/2. Table 3.3 shows the different weave structure and their presentation.

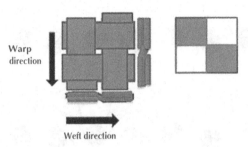

Warp
direction

Weft direction

FIGURE 3.26   Plain weave.

While Figure 3.27 illustrates the interlacement of the weft and warp yarns for the various designs in the fabric cross-section per repeat.

The yarn interlacing will make the length of the yarns higher than their length in the formed fabric which is due to the yarn crimp, either in the weft direction or warp direction. The numbers of the warps per cm as well as the number of weft per cm predetermine several physical and mechanical properties of the fabric, while the number of yarn intersections per repeat is responsible for the main physical and mechanical properties of the fabric. From Figure 3.27, it can be seen that the plain weave has the highest number of weft and warp yarn intersections for the same repeat for all the fabric designs.

The fabric surface topography affects the interfacial strength of the fabric in the matrix. Figure 3.28 illustrates that, if the fabric design has a float as in the case of the twill or satin weave, it will have higher space area between the yarns to allow more polymer and reduce fiber volume fraction, than in the plain weaves, changing the mechanical properties of the final composite.

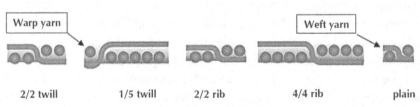

Warp yarn   Weft yarn

2/2 twill   1/5 twill   2/2 rib   4/4 rib   plain

FIGURE 3.27   Warp and weft yarns intersections in different fabric structures per repeat.

TABLE 3.3   Fabric Design for Laminates

Type of fabric design	Presentation	Type of fabric design	Presentation
Plain weave		2/2 rib weft weave	
2x2 mat weave		2/2 rib warp weave	
1/3 twill weave		Compound design	
1/4 sateen weave		Crow fort satin weave	
2/1 twill		2/2 twill	

Figure 3.29 demonstrates how the macro-topography of the fabric of the same design but different yarn count looks unlike. This is due to the fact that higher crimp is noticed when using course yarns.

Referring to spun yarns, the fiber protruding outside the yarn (Hairiness) will also change the surface topography of weave used. Yarn irregularities will cause macro-topographic irregularities that may have the impact on the mechanism of formation of micro cracks.

3.6.2   FABRIC TECHNICAL PARAMETERS OF WOVEN FABRIC

In order to specify the specifications of the fabric used as laminate in the composite material, several parameters should be calculated.

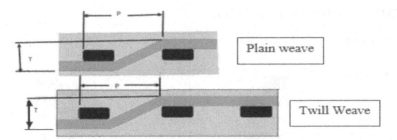

FIGURE 3.28   Fiber polymer composite for plain and twill weaves.

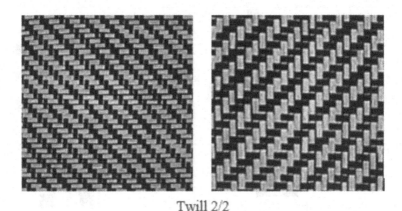

Twill 2/2

FIGURE 3.29   Twill fabric.

3.6.2.1   Yarn Diameter

The yarn diameter d is given by,

$$d = 0.00357 \ (tex/\phi \ \rho)^{0.5} \tag{1}$$

where: d – yarn diameter cm; tex – yarn linear density "g/km"; ρ – fiber density g/cm^3; ϕ = yarn packing coefficient.

The value of the packing coefficient is a function of the type of spinning system twist factor and the fiber properties, its value approximately taken as 0.67 [32].

3.6.2.2 Yarn Count

The linear density of yarn is expressed by indirect system "length/weight" or direct system "weight/length." The "ISO" system is the direct system defined by *tex*, denier.

Denier = (weight of yarn of length L in "g")/
(length of yarn in 9 "km" length unit) (2)

tex = (weight of yarn of length L in "g"/
length of yarn in "km" length unit) (3)

The value of yarn *"tex"* which has length of one "km" and weight one gram will be one *tex*. One *denier is* equal to *9 tex*.

3.6.2.3 Fabric Crimp

Weft crimp is due to waviness caused by weft yarns interlacements over the warp yarns to form the fabric and is equal to:

$$C\% = [(l_{yarn} - l_{fabric}) / l_{fabric}] \times 100 \qquad (4)$$

The value of the weft or warp crimp will depend on the following parameters:
- Fiber properties
- Yarn count
- Yarn twist
- Spinning system (Ring, Open End, Friction, Compact spinning)
- Type of yarn (staple, continuous filament, flat, etc.)
- Number of weft/cm
- Number of warp/cm
- Fabric tightness
- Fabric design

The crimp value affects most of the physical and mechanical fabric properties.

3.6.2.4 Fabric Areal Density

The fabric areal density "g/m²" has consequence on the different composite mechanical properties hence it determines the fiber volume fraction.

Weight of fabric gm/m² = Weight of weft /m² + Weight of warp /m²

$$\textit{Weight of weft "g/m²"} = \textit{(Ends per cm) } tex_{weft}\textit{ (1 + crimp }_{weft}\textit{%)/10} \quad (5)$$

$$\textit{Weight of warp "g/m²"} = \textit{(Warps per cm) } tex_{warp}\textit{ (1 + crimp }_{warp}\textit{%)/10} \quad (6)$$

where, tex_{weft} weft yarn count, tex_{warp} warp yarn count.

3.6.2.5 Fabric Fractional Cover

In fabrics constructed from yarns, the area covered may be considered as the *Fractional* of the total fabric area that "covered" by one-component yarns.

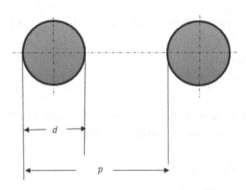

$$p = \left(\frac{threads}{cm}\right)^{-1}$$

FIGURE 3.30 Yarn of circular cross section.

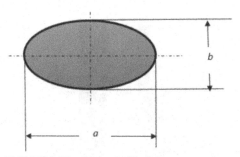

FIGURE 3.31   Yarn of elliptical cross section.

Assuming the yarns of circular cross section [33] Figure 3.30, and then the area covered by the yarns in unit cell will be:

Fractional Cover = 0.00444 (tex /fiber density)$^{0.5}$
$\qquad\qquad$ *× (number of threads per cm) %* $\qquad\qquad$ (7)

For yarn of elliptical cross section, Figure 3.31, with ratio of (a/b) = R, the fractional cover of the yarns per unit cell (FC %) will be again:

Fractional Cover = 0.00444 (R. (tex /fiber density))$^{0.5}$
$\qquad\qquad$ *× (number of threads per cm) %* $\qquad\qquad$ (8)

Assuming the thickness of the laminate equals to the continuous filaments yarn thickness, the volume fraction will be approximately equal to:

Fiber volume fraction = FC %.

3.6.2.6   Total Fabric Cover

The definition of "Cover factor" indicates the extent to which the area of a fabric is covered by one set of threads.

For any fabric there are two *Fractional* cover factors: the warp *Fractional* cover factor and the weft *Fractional* cover factor.

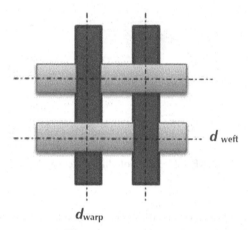

FIGURE 3.32 Fabric cell.

Fabric cover factor is sum of the weft *Fractional* cover and the warp *Fractional* cover subtracting the common areas as illustrated in Figure 3.32. Then the total fabric cover area by weft and warp yarns can be given by:

The total fabric cover factor of the fabric $C_t = C1 + C2 - C1C2$ (9)

Fractional Cover of the weft C1; *Fractional* Cover factor of the warp C2.

The total cover factor gives percentage covered by fibers and the rest will be filled by the matrix. When ignoring that, matrix polymer may penetrate inside the yarns. The percentage of fiber volume fraction in laminate:

$V_f = $ *(Volume of the fibers/Volume of the laminate)* $\times 100$ (10)

V_f *is approximately equal to* $= (C_t) \times 100$ (11)

It was found that the fabric cover factor influences the following properties:
- fabric porosity
- fabric strength and elongation
- fabric stiffness
- abrasion resistance

For different fabric designs, the value of the cover factor for loose structures varied between 25–50%, for regular structures 50–70%, and for tight structures 75–100%.

3.6.2.7 Weave Areal Density

Fabric density indicates the number of warp yarns and weft yarns in the unit area.

$$Warp\ density = ends\ per\ cm\ (tex_{warp})^{0.5}\ k^{0.5} \tag{12}$$

$$Weft\ density = picks\ per\ cm\ (tex_{weft})^{0.5}\ k^{0.5} \tag{13}$$

The value of "k" is equal to 1 for plain weave, twill 1/2 (0.87), twill 1/3 (0.77), satin weave (0.69).

Fabric areal density g/cm² = Warp weight/cm² + Weft weight/cm² (14)

$$Warp\ weight\ g/cm^2 = (ends\ per\ cm)\ (tex_{warp})\ (1 + C_{warp})\ 10^{-1} \tag{15}$$

$$Weft\ weight\ g/cm^2 = (picks\ per\ cm)\ (tex_{weft})\ (1 + C_{weft})\ 10^{-1} \tag{16}$$

where, tex_{warp} – Warp count; C_{warp} – Contraction of warp yarns; tex_{weft} – Weft count; C_{weft} – Contraction of weft yarns.

3.6.2.8 Fabric Specific Volume

The apparent specific volume of fabric, V_{fabric}, is calculated as:

$$V_{fabric} = Fabric\ thickness\ "cm"/Fabric\ mass\ per\ unit\ area\ (g/cm^2) \tag{17}$$

The fabric packing factor ϕ is given by:

$$\phi = (V_{fiber}/V_{fabric}) \tag{18}$$

3.6.2.9   Weave Factor

It is calculated from the number of the interlacements of warp and weft in a given repeat. It is also equal to an average float and is expressed as:

WF = Number of threads per repeat /
$\qquad$ *Number of intersections per repeat* $\qquad$ (19)

The weave factor is considered in direction of weft or warp. The resultant weave factor is expressed as the product of the weave factor in both directions. For plain weave WF_{weft} is equal to 1, for twill 2/1 is 1.5, and for warp rib 2/2 is also 1.

The fabrics with lower value of WF in any direction will have higher strength and lower extension. This will change the mechanical properties of the laminates, assuming the same yarns are used in both weft and warp directions [33, 34]. For compound structures of weave design "WF" will be:

WF = Number of threads per repeat/
$\qquad$ *Number of intersections per repeat* $\qquad$ (20)

3.6.2.10   Woven Fabric Jamming

Woven fabric consists of set of interlaced warp and weft yarns in a certain manner, according to the fabric design. The mobility of the yarns depends on the number of warp and weft yarns per cm and their diameters. Under the loading, they may move till be jammed. In jammed structures, the warp and weft yarns will have minimum thread spacing [33], so they have no mobility within the structure as they are in the intimate contact with each other. Applying Pierce's Geometrical model of woven structures, the simple relation between weft and warp diameter d_1, d_2, P_1 weft thread spacing, and P_2 warp thread spacing for a fabric being jammed in both directions, is:

$$(1 - (P_1/d_1)^2)^{0.5} + (1 - (P_2/d_2)^2)^{0.5} = 1 \qquad (21)$$

3.6.2.11   Effect of Woven Fabric Design on Some Mechanical Properties

The fabric design affects both the mechanical and physical properties of the fabric either in weft or warp directions. Generally, the number of the intersections per repeat is linearly proportional to the fabric strength and the shear properties of the fabric. Figure 3.33 illustrates the shear angle, shear force, shear stress and shear modulus, is found to be highly

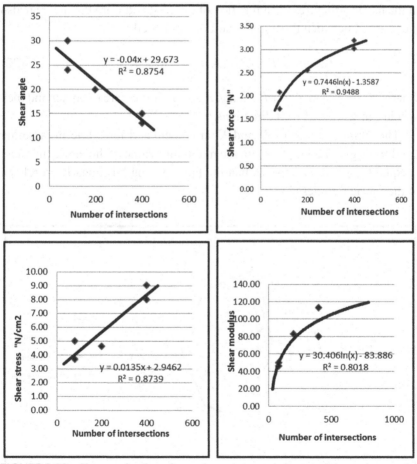

FIGURE 3.33   Shear angle, shear force, shear stress, and shear modulus versus number of intersections.

correlated to the number of the intersection in spite of the fabric design [33, 34].

The number of intersection also affect the fabric tightness [35], in the same time, the fabric tightness affects all the mechanical properties of the fabric and the air permeability and porosity. The effect of the fabric design on the various fabric properties are ranked and illustrates in Figure 3.34. The fabric design with higher number of intersections per repeat will have higher strength, Young's modulus, bending stiffness, shear modulus, crimp, and lower elongation, porosity.

Fabric tightness is a measure of how the weft and warp yarns are packed in the fabric structures it's a function of the weft and warp diameter an also the spacing between them in the unit of fabric cell.

$$Tightness\ T = (K_1 + K_2)_{actual}/(K_1 + K_2)_{limit} \tag{22}$$

where: $K_1 = (d_1/P_1)$, $K_2 = (d_2/P_2)$ and d_1, d_2, P_1 and P_2 are diameter and pick spacing in warp and weft directions.

The limited values of K_1 and K_2 are calculated for fabric in jamming conditions given by Eq. (21). In the maximum value of the fabric tightness is equal to one. It was proved that the fabric strength is linearly correlated to the fabric tightness [33].

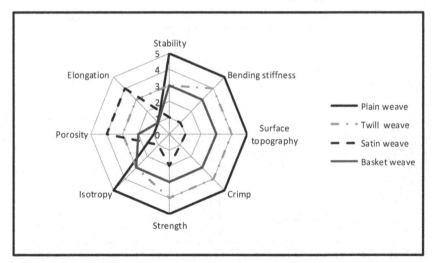

FIGURE 3.34 Ranking of the properties of laminate related to fabric design.

3.6.3 KNITTED FABRIC

The knitted fabric can be used for the formation of preforms of the composite, because of its simplicity, low cost, can give neat shapes in one step, rapid manufacturing of the prototype reinforcements, controlled anisotropy. The fabric design of jersey knit allow high fabric stretch in course direction, it may reach 100%, helping in the formation of complex preform shapes. The structure of the rib knit as shown in Figure 3.10 where more weft yarn can be used and the fabric is more bulky. The rib knit can be used to form 3-D shapes on the surfaces of the fabric.

3.6.3.1 Knitted Fabric Technical Parameters

The basic calculations of the knitted fabric which consequence in its physical and mechanical properties are:

3.6.3.1.1 Weft Knitted Structures

- Courses/cm "cpc" = k_c/l
- Wales/cm "wpc" = k_w/l
- Stitch density per cm^2 "s" = (cpc).(wpc) = k_s/l^2
- Shape factor = (cpc)/(wpc) = k_c/k_w
- Fabric tightness

The fabric tightness factor K of a knitted fabric is defined as the ratio of the fabric area covered by the yarn to the total fabric area.

$$Fabric\ tightness\ "K" = (tex)^{0.5}/l \qquad (23)$$

where: l the loop length "mm"; k_c, k_w and k_s are constants depended on the materials of the used yarns and condition of the fabric relaxations [36, 37].

The fabric tightness will affect the spirality angle, which generally decreases as the tightness of the fabric increases. The value of the fabric tightness has a mean value of 1.47 for single jersey and 1.5 for weft knitted structure. The value of fabric areal density is given by:

$$Fabric\ areal\ density\ "g/m^2" = 0.01\ s\ l\ tex \qquad (24)$$

3.6.3.1.2 Warp Knitted Structures

- Tightness factor: The tightness factor of two guide bars (front f and back b) is given by:

$$K = (tex_f)^{0.5}/l_f + (tex_b)^{0.5}/l_b \qquad (25)$$

- Fabric areal density of two guide bars:

$$\textit{Fabric areal density "g/m}^2\textit{"} = s((l_f.tex_f) + (l_b.tex_b)) \times 10^{-2} \qquad (26)$$

The fabric various parameters will affect the physical and mechanical properties of the preform [31] and therefore, should be chosen to fulfill the requirements of the composite specifications.

3.6.4 Braided Preform

3.6.4.1 Geometrical Parameters of Braided Fabric

The braided weave is a biaxial flat braid which consists of number of yarns interlacing together. The simple geometrical parameters of the proposed model are shown in Figure 3.35.

The equations for tubular braids were modified by replacing the braid perimeter by the braid width to get the suitable model, which is applicable for flat braids [38].

3.6.4.2 Braid Angle α

The key design parameters of the braid are: diameter of the yarn (d), the braid width (W) and the braid angle (α), Figure 3.35. A unit cell of the braid is formed by the interlacements of yarns. Thus, the braid angle (α) is the most important parameter [39], which is defined as the angle between the yarn axes and braid axes. Accordingly, the braid angle (α) can be calculated based on the braid pitch (P), as shown in the following equation:

$$Sin\ \alpha = W/(N.P) \qquad (27)$$

where, N is number of yarns.

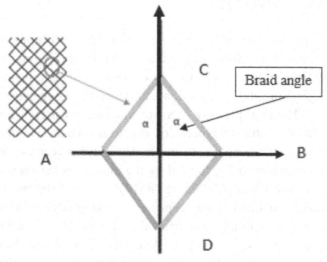

FIGURE 3.35   Biaxial braid.

3.6.4.3   Braid Jamming Condition

Under low-loading conditions, the braids go through the geometric transition, i.e. occurrence of yarn reorientation, and the braid angle (α) decreases resulting in an increase in length and decrease in braid width. According to the above justification, the braid angle continues to decrease until the yarns become tightly packed and the structure is jammed such that no further movement of yarns can take place. Under elongation of the braided fabric, the movement of yarns towards the braid axis occurs under low loading conditions without strain in yarns, resulting in decreasing of braid angle "α " and braid width. The increasing of load leads to increase in the compression of the braid in the width direction, and decreasing in braid angle a matter that cause yarns to contact each other. Further increasing of load causes the yarns to extend and leads to further yarn packing (jammed condition is reached). The jamming or locking of structure will lead to minimum possible spaces between the constituent yarns, and the new average spacing between the yarns after jamming in a unit cell.

3.7   FABRIC POROSITY

Porosity strongly determines important properties of fibrous materials, such as resin management properties, air permeability, and mechanical properties. Large surface area to volume ratio, (as a result of high porosity) gives the ability to facilitate interfacial bonding between fibers and matrix. The total porosity of a woven fabric consists of three components: (a) the intra-fiber porosity, due to the voids within the fiber itself, (b) the inter-fiber porosity, due to the voids between the fibers, and (c) the inter yarn porosity, formed by the interstices between the yarns [41]. The Resin Transfer Molding (RTM) is one of the most promising technology available today capable of making large complex three-dimensional part with high mechanical performance, tight dimensional tolerance and high surface finish. Whatever is the fabric design, it is very important to allow the resin to flow at suitable speed through all the parts of the structure without high resistance to the resin flow, leaving a minimum percentage of voids when the mold is closed and the preform reaches the expected fiber volume fraction. If the fibers are too compacted or their content is excessive, there is no sufficient space for the passage of the resin and the filling time becomes longer. An increase with respect to this value may cause a bad distribution of the resin and a dramatic drop in the mechanical properties of the manufactured component. Micro-voids may cause a micro crakes, leading to the delamination and early failure of the composite [42]. The porosity of woven fabric is well defined due to the pattern of the fabric with a predetermined yarn interlacements, which indicated that the higher the number of interlacements of the warp and weft yarns, the lower is the porosity. While non-woven fabric fibrous material may contain three kinds of pores: closed pores, blind pores and through pores [43]. Generally, the higher fiber volume fraction – the lower is the porosity of the fibrous structure. The porosity can be defined by several parameters, average diameter of pores and distribution of pore's diameters, total number of pores. The pore distribution is an important parameter in defining the rate of flow of the resin, while the size of pores determines the percentage of voids in the final composite.

KEYWORDS

- **2-D fabric**
- **braided fabric**
- **classification of composite**
- **cover factor**
- **design of laminates**
- **fabric jamming**
- **fabric porosity**
- **fabric specific volume**
- **knitting fabric**
- **nonwoven fabric**
- **triaxial fabric**

REFERENCES

1. El Messiry M.; El Deeb, R. Study Wheat Straw Composite Properties Reinforced by Animal Glues as Eco-Composite Materials, The 2nd Conference for Industrial Textile Researches, National Research Centre (NRC), Cairo, Egypt, 2013.
2. Botros, M. Development of new generation coupling agents for wood-plastic composites. [Online] http://www.lyondellbasell.com/techlit/techlit/Tech%20Topics/Equistar%20Industry%20Papers/CouplingAgents%20for%20Wood-Plastic.pdf (accessed May 5, 2016).
3. Shanmugasundaram, O. L. Green composites, Manufacturing Techniques & Applications. Indian J. T. October, 2009.
4. Bekaert Co. http://www.bekaert.com/ (accessed May 5, 2016).
5. Pan, X.; Sano, Y. Fractionation of wheat straw by atmospheric acetic acid process, Bioresour. Technol. 2005, 96(11), 1256–1263.
6. Mansour M. R.; Hamble A.; Azaman M. D. Material selection of thermoplastic matrix for hybrid natural fiber/glass fiber polymer composites using analytic hierarchy process method. Proceedings of the International Symposium on the Analytic Hierarchy Process, 2013.
7. Saba, N.; Tahir, P.; Jawaid, M. A Review on Potentiality of Nano Filler/Natural Fiber Filled. Polym. 2014, 6, 2247–2273.
8. Cherouat, A.; Borouchaki, H. Present State of the Art of Composite Fabric Forming: Geometrical and Mechanical Approaches, Materials. [Online] 2009, 2, 1835–1857. file:///C:/Users/magdy/Downloads/materials-02–01835%20(6).pdf (accessed May 6, 2016).

9. Wencheng, L. Principles for determining material allowable and design allowable values of composite aircraft structures, Procedia Eng. 2011, 17, 279–285 file:///C:/Users/magdy/Downloads/materials-02–01835%20(5).pdf (accessed May 5.2016).

10. NASA, Marshall Space Flight Center (MSFC), Design and manufacturing guideline for aerospace composites. GUIDELINE NO. GD-ED-2205 http://oce.jpl.nasa.gov/practices/2205.pdf (accessed May 5. 2016).

11. STIG, F. 3D-woven Reinforcement in Composites, Doctoral Thesis, KTH School of Engineering Sciences, Stockholm, Sweden 2012.

12. STIG, F. An Introduction to the Mechanics of 3D-Woven Fiber Reinforced Composites, PhD. Thesis, KTH School of Engineering Sciences, Stockholm, Sweden 2009.

13. Unal, U. P. G. Classification of 3-D woven fabrics, Woven Fabrics, Chapter 4, Edited by Han-Yong Jeon, Publisher: InTech, Chapters published, 2012.

14. Cherouat, A.; Bourouchaki, H. Numerical Tools for Composite Woven Fabric Preforming. Adv. Mater. Sci. Eng. [Online] 2013, Vol. 2013, 1–18 http://www.hindawi.com/journals/amse/2013/709495/ (accessed May 5.2016).

15. Raju, J.; Manjusha, S.; Duragkar, N.; Jagannathen, C.; Manjunatha, M. Prediction of onset of mode I delamination growth under a tensile spectrum load, J. Mater. Sci. Res. 2014, 3(2), 45–51.

16. Frank L. Scardino, F. L.; Ko, F. Triaxial Woven Fabrics, Part I: Behavior Under Tensile, Shear, and Burst Deformation. Text. Res. J. 1981, 51(2), 80–89.

17. Triaxial structures, Inc., http://www.triaxial.us/index.php (accessed April 8, 2016).

18. Sengupta, S.; Chattopadhyay, S.; Samajwadi, S.; Day, A. Use of jute needle-punched nonwoven fabric as reinforcement in composite. IJFTR 2008, 33, 37–44.

19. Roy, A. N.; Ray, P. Optimization of Jute Needle-Punched Nonwoven Fabric Properties: Part II—Some Mechanical and Functional Properties. J.NAT. FIBERS 2009, 6(4), 303–318.

20. Maity, S.; Singha, K.; Gon, D. P.; Paul, P.; Singha, M. A Review on Jute Nonwovens: Manufacturing, Properties and Applications. IJTS (Online) 2012, 1(5), 36–43 file:///C:/Users/magdy/Downloads/10.5923.j.textile.20120105.02%20(1).pdf (accessed May 1, 2016)

21. Drechsler, K. Advanced textile structural composites-needs and current developments. The 5th International conf. on textile composites, Leuven, Belgium, 2000.

22. Potluri, P.; Rawal, A.; Rivaldi, M.; Porat, I.; Geometrical modeling and control of a triaxial braiding machine for producing 3-D preforms. Composites: Part A 2003, 34, 481–492.

23. Branscomb, D. New Directions in Braiding, Royall Broughton, New Directions in Braiding. JEFF 2013, 8(2), 11–24.

24. Bogdanovich, A.; Mungalov, D. Recent advancements in manufacturing 3-D braided preforms and composites. Proceedings of ACUN-4 composite systems—macro composites, micro composites, nanocomposites, University of New South Wales, Sydney, Australia, Bandyopadhyay, S. (Ed.), 2002.

25. Kohlman, L. W. Evaluation of test methods for triaxial braid composites and the development of a large multiaxial test frame for validation using braided tube specimens. PhD Theses, the Graduate Faculty of The University of Akron, USA, 2012.

26. Corne, D.; Sun, D.; Stylios, G. Revolutionizing the design of 3-D textile composites. [Online], http://cdi.hw.ac.uk/3d-textile-composites/ (accessed May 5, 2016).

27. Khokar, N.; Winberg, F. 3-D Fabrics for Composites: Limitless Opportunities, Limited Processes, 6th World Conference on 3-D Fabrics and their Applications, Raleigh, NC, USA, 2015.

28. Hearle, J. W. S.; Chen, X. 3-D woven preforms and properties for textile composites. [Online] http://iccm-central.org/Proceedings/ICCM17proceedings/Themes/Materials/3D%20TEXTILES%20&%20COMP/D1.13%20Hearle.pdf (accessed May 1, 2016)

29. Cherouat, A.; Borouchaki, H. Present state of the art of composite fabric forming: geometrical and mechanical approaches. Materials [Online] 2009, 2, 1835–1857. file:///C:/Users/magdy/Downloads/materials-02–01835%20(6).pdf (accessed May 6, 2016).

30. Abounaim, Md. Development of flat knitted spacer fabrics for composites using hybrid yarns and investigation of two-dimensional mechanical properties. Text. Res. J. 2009, 79(7), 596–610.

31. Armakan, D. M.; Roye, A. A study on the compression behavior of spacer fabrics designed for concreate applications. Fiber Polym. 2009, 10(1), 116–123.

32. Kilic, M.; Buyukayraktat, R.; Kilic, G.; Aydin, S.; Easki, N. Comparing the packing density of yarns spun by ring, compact and vortex spinning systems using image analysis methods. IJFTR, 2014, 39, 351–357.

33. Behera, B. K.; Militky, J.; Mishra, R.; Kremenakova, D. Modeling of Woven Fabrics Geometry and Properties, Woven Fabrics. Edited by Han-Yong Jeon, Intech, May 16, 2012.

34. El Messiry, M.; El-Tarfawy, S. Effect Of Fabric Properties On Yarn Pulling Force For Stab Resistance Body Armour, Sixth World Conference On 3-D Fabrics and Their Applications. NCSU, Raleigh, NC, USA, 2015.

35. Seyam, A.; El-Shiekh, A. Mechanics of Woven Fabrics Part IV: Critical Review of Fabric Degree of Tightness and Its Applications. Text. Res. J. 1994, 64(11), 653–662.

36. Kumar, V.; Asampth, V.; Vigneswaran, C. study on geometric and dimensional properties of double pique knitted fabric using cotton sheath elastomeric core spun yarn, JTATM, 2014, 8, 1–13.

37. Gravas, E.; Kiekens, P.; Van Langenhove, L. Predicting fabric weight per unit area of single- and double-knitted structures using appropriate software. AUTEX Res. J. 2006, 6(4), 223–237.

38. Byun, J. The Analytical Characterization of 2-D Braided Textile Composites. Compos. Sci. Technol. 2000, 60(5), 705–716.

39. Rawal, A.; Rajesh Kumar, R.; Saraswat, H. Tensile mechanics of braided sutures. Text. Res. J. 2012, 82(16), 1703–1710.

40. Potluri, P.; Rawal, A.; Rivaldi, M.; Porat, I. Geometrical modeling and control of a triaxial braiding machine for producing 3-D preforms. Composites Part A 2003, 34, 481–492.

41. El Messiry, M. Theoretical analysis of natural fiber volume fraction of reinforced composites. A. E. J. 2013, 52(3), 301–306.

42. Laurenzi, S; Marchetti, M. Advanced composite materials by resin transfer molding for aerospace applications, chapter 10, Composites and their properties. Edited by Ning Hu, Intech, 2012.

43. Rasi, M. Permeability Properties of Paper Materials. PhD Theses, Department of Physics, University of Jyväskylä, 2013.

CHAPTER 4

TEXTILE REINFORCEMENT MODIFICATION AND MATRIX MATERIALIZATION

CONTENTS

4.1   INTRODUCTION

Usually, the natural fibers cannot be used directly in the manufacturing of natural fiber composites (NFPC) because of the difference in the surface properties of the two materials. The properties of the NFPC depend on the interfacial strength. Low value of interfacial strength will lead to propagation of the micro crack at the reinforcement surface separating the matrix, thus the composite loses its integrity. The most serious concern with natural fibers is their hydrophilic nature due to the presence of pendant hydroxyl and polar groups in various constituents, which can result in poor adhesion between fibers and hydrophobic matrix polymers [1].

Several approaches of surface modification are used to increase the natural fibers–matrix interfacial strength before the injection of a polymer. Physical and chemical methods optimize the interface between the fiber and the matrix. Coupling agent commonly improve the degree of cross-linking in the interface region and offer a perfect bonding [2, 3].

4.2   METHODS FOR FIBER SURFACE TREATMENT

The natural fibers can be used to reinforce both thermosetting and thermoplastic matrices. Thermosetting resins, such as epoxy, polyester, polyurethane, phenolic, etc. are commonly used in natural fiber composites, when there is the requirement for higher performance applications. The chemical composition of the natural fibers are cellulose (α-cellulose) hemicelluloses, lignin, pectin and waxes. Their percentage depends on the type of fibers. Table 4.1 gives the analysis of the

TABLE 4.1   Chemical Composition of Some Natural Fibers [4–6]

Fiber	Cellulose %	Hemicellulose %	Lignin %	Waxes %
Abaca	56–63	20–25	7–9	3
Alfa	45.4	38.5	14.9	2
Bagasse	55.2	16.8	25.3	–
Bamboo	26–43	30	21–31	–
Banana	63–64	19	5	–
Coir	32–43	0.15–0.25	40–45	–
Cotton	85–90	5.7	–	0.6
Curaua	73.6	9.9	7.5	–
Flax	71	18.6–20.6	2.2	1.5
Hemp	68	15	10	0.8
Isora	74	10	14	1.09
Jute	61–71	14–20	12–13	0.5
Kenaf	72	20.3	9	–
Nettle	60–86	10–20	5–10%	4
Oil Palm	42–65	6–32	22–29	–
Pineapple	81	12–16	12.7	–
Ramie	68.6–76.2	13–16	0.6–0.7	0.3
Sisal	65	12	9.9	2
Straw (Wheat)	38.45	15–31	12–20	–

fiber chemical composition of some natural fibers [4–6]. It is noticeable that the percentage of cellulose is varied between 30 and 86%. Most methods for the modification of natural fibers depend on the cellulose percentage in the treated fiber. Several researchers suggest various methods to improve the fiber surface to make it suitable for good interfacial fiber/matrix bonding reinforce with thermoset, thermoplastic, biodegradable plastics, and natural rubber. The different surface chemical modifications of natural fibers, such as alkali treatment, silane treatment improve the adhesion between the matrix and the fibers. A third component, called compatibilizer, has to be used so that the fiber's surface is modified prior to the preparation of the composite [3–6]. Treatments like isocyanate treatment, latex coating, permanganate treatment, acetylation, monomer grafting under UV radiation, etc., have achieved various levels of success in improving fiber strength and fiber/matrix adhesion in natural fiber composites [7].

The potential of use cellulose nanofiber in nanocomposite shows the increasing interest, especially in bio-nanocomposites, that requires good bonding and distribution between cellulose nanoparticles and the matrix to improve their interfacial properties.

The methods for modifying the natural fibers to improve the fiber-matrix interface can be divided as given in the following subsections [8, 9].

4.2.1 PHYSICAL METHODS FOR FIBER SURFACE TREATMENT

The physical methods will not alter the chemical composition of the fibers but change both the structure and the surface morphology of the fibers. This includes stretching, calendaring, thermos-treatment or cold plasma treatment to achieve modification of the surface properties of the fiber to modify the bonding with the matrix. The application of hybrid yarns changes the structural and surface properties of the yarns and thereby, influences the mechanical bonding to polymers [11]. On modern spinning machines, several methods were introduced to change the geometry of the yarns, such as the formation of slub structures for alternate the surface morphology of the yarns providing a better bonding between the matrix and the yarns.

Plasma treatment methods are known to be effective for "non-active" polymer substrates as polystyrene, polyethylene, polypropylene, etc. They are successfully used for cellulose-fiber modification to decrease the melt viscosity of cellulose-polyethylene composites and improve mechanical properties of cellulose-polypropylene composites [12].

4.2.1.1   Plasma Treatment

Plasma is a "dry" technology and is intrinsically ecological and environment friendly. It can be used to modify the properties of the surfaces in a wide range of textiles fibers without change to the bulk properties. Low-temperature plasma treatment can be an alternative to traditional wet processes in textile. Compared with current standard finishing processes, plasmas have the fundamental advantage of reducing the usage of chemicals, water and energy [9, 13]. Moreover, fiber surfaces are etched consequently the changes in fiber diameter and surface roughness.

The plasma industry has developed equipment configurations that run a hybrid corona/DBD plasma type that is universally called "Corona." Plasma technology performed under atmospheric pressure or under reduced pressure (depending on the special needs) leads to a variety of processes to modify fiber to fulfill additional highly desirable requirements. Plasma technology is suitable to modify the chemical structure as well as the topography of the surface of natural or man-made fibers [12–22]. Plasma resource could be classified accordingly to source: Low pressure and atmospheric pressure. The atmospheric pressure is most commonly used with textile processing and less expensive. Corona treatment is one of the most interesting techniques for surface oxidation activation. It is a process that changes the surface energy of the cellulose fibers and, in case of wood surface, increases the amount of aldehyde groups. Depending on the type and nature of the gases used, a variety of surface modification could be achieved by using cold plasma treatment which will in turn introduce surface crosslinking while surface energy could be increased or decreased and reactive free radicals and groups could be produced [13].

Plasma treatment enhances inter yarn and inter fiber friction due to etching effect. Rougher surface of individual fibers increases the friction between fibers within the yarn and also between the yarns, and improve the interfacial strength [14]. Some results of sisal fiber treatment by cold plasma indicate that the improvement of its mechanical properties depends on the time of exposition to treatment. Study of the effect of the plasma treatment on the flax fiber shows no significant increase in the fiber tenacity but an increase in the breaking elongation by about 60%.

The various results are noticed due to the difference of the source of the bast fibers hence the diameter of the fibers was found to be a decisive parameter that determines the effect of the plasma treatment [15]. The analysis of topography, as illustrated in Table 4.2 of fiber by SEM, shows the etching of the surface of the fibers, causing the increase in the interfacial force between the fiber and matrix [20].

The plasma treatment of flax fiber indicates a decrease of the Young's modulus and strain without significant reduction of the strength.

4.2.2 CHEMICAL METHODS FOR FIBER SURFACE TREATMENT

To develop composites with good properties, it is necessary to improve the fiber-matrix interface and reduce moisture absorption. Chemical treatments target to improve interfacial adhesion necessary to overcome the

TABLE 4.2   SME Photo of Fibers After Plasma Treatment

Fiber type		Fiber type	
Nylon		Wool	
Polyester		Cotton	

weak bonding, reduce moisture absorption, and improve the mechanical properties of the fibers [2–19]. The mechanism of the improving the interfacial strength between the incompatible fibers and matrix materials may be achieved through:

- Reducing the cellulose hydroxyl groups
- Grafting
- Cross-linking the cellulose of the fibers with matrix polymer
- Thoroughly cleaning the fiber surface by bleaching

The most optimum method depends on the type of the fiber and polymer of the matrix used for manufacturing the composite [21–23].

The different methods of chemical pre-treatment of the natural fibers, as well as their effects, are given in Table 4.3. The effects of the chemical modifications were investigated [23–27] and shown their impact on the performance of natural fibers and fiber reinforced composites. The different chemical modifications of natural fibers surface achieved various levels of success in improving fiber strength and fiber-matrix adhesion in NFPC. However, in some cases, the main chemical bonding theory alone is not sufficient: hence the consideration of other concepts appears to be necessary, which include the morphology of the interface, the acid-base reactions at the interface, surface energy and the wetting phenomena [28]. Pretreatment methods are of different effectiveness for the interfacial adhesion between matrix and fiber, depending on the type of fiber and the polymer used.

4.2.2.1 Alkaline Treatment

The alkaline treatment is the most popular treatment for the bast fibers to improve their properties. Mercerization leads to the increase in the amount of amorphous cellulose at the expense of crystalline cellulose. The important modification expected here is the removal of hydrogen bonding in the network structure. The following reaction proceeds as a result of treatment with alkalis:

$$\text{Fiber–OH} + \text{NaOH} \quad \rightarrow \quad \text{Fiber–O–Na}^+ + \text{H}_2\text{O}$$

As a result of sodium hydroxide penetration into crystalline regions of parent cellulose (cellulose I), alkali cellulose is formed. Then, after washing

TABLE 4.3 Methods of Chemical Pre-Treatment of Natural Fibers [21–28]

Chemical	Reaction	Effect
Alkaline treatment	Removes wax, lignin, breaking the hydrogen bonding	Change of orientation of crystalline cellulose, increase surface roughness
Silane (SiH4) treatment	Reduces cellulose hydroxyl group, reacts with the hydroxyl group	Used as coupling agents, reduces the fiber swelling
Benzoylation treatment	Benzoyl treatment	Decrease water absorption
Acylation treatment	Acrylic acid ($CH_2=CHCOOH$) grafting	Cause plasticization of cellulosic fiber, it becomes hydrophobic
Coupling agent	Develops a highly cross-linked interphase region, with a modulus intermediate between substrate and the polymer	Esterification of cellulose fiber
Peroxide treatment	Benzoyl peroxide, Dicumyl peroxide are used with cellulosic fibers for surface modification after alkali pre-treatment	Treatment of the fibers after alkaline pre-treatment.
Sodium chlorite treatment	Bleaching of the cellulosic fibers with sodium chlorite will clean the surface of the fibers to allow better adhesion with the matrix	Clean the area of the fibers, makes its surface rough.
Permanganate treatment	Potassium permanganate solution in acetone in different concentration	Treatment of the fibers after alkaline pre-treatment.
Acetylation treatment	Esterification method causes plasticization of cellulose fibers, modifying the properties of polymer so that they become hydrophobic	Increase in hydrophobic character
Acylation	Treatment with acrylic acid	Graft polymerization modify natural fiber

out unreacted NaOH, the formation of regenerated cellulose (cellulose II) takes place. The alkaline treatment has two effects on the fiber:

(1) It increases surface roughness resulting in better mechanical interlocking;

(2) It increases the amount of cellulose exposed on the fiber surface, thus increasing the number of possible reaction sites [29].

SEM photos for untreated and treated jute fibers are illustrated in Figure 4.1 which points out that NaHO treatment affects the morphology of the fibers.

Consequently, alkaline treatment has a permanent effect on the mechanical behavior of jute fibers, especially on fiber's strength and stiffness [30–34]. The mercerization of ramie yarn has a considerable effect on crosslinking properties, such as the degree of cross linking, tensile strength, and elongation [35]. The medium concentration of (NaOH) has reoriented the fibers and, consequently improved their crystallinity and strength, NaOH concentration of 10% gave the best effect [36]. The degumming process is necessary for improving the textile properties of jute fiber, accordingly fineness 2.02 *tex* was obtained [37]. Changes, occurring in jute fiber-PP composites when fibers are pretreated with 2% concentration of a NaOH for 24 hour, were investigated [38] and indicated the improve in the mechanical properties due to good adhesion between natural fibers and polypropylene (PP) matrix. Jute fibers, when alkali-treated with 2% concentration for 24 hr. showed best improvement in tensile strength by 40% and modulus by 9%. Moisture absorption for jute composites was 50% lower than for the untreated fiber composites. Also, it was observed that the alkali treatment produced a drop in both tensile

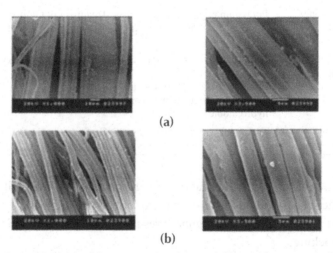

(a)

(b)

FIGURE 4.1 SEM photos for untreated and treated fibers: (a) untreated, (b) treated (NaOH concentration 8%, immersion time 30 min).

strength and Young's modulus of the fibers [39]. This was attributed to the damage induced to the cell walls and the excessive extraction of lignin and hemicellulose, which play a significant role in the structure of the fibers. The tensile properties of the alkali treated woven jute natural fibers reinforced Hybrid Composites [40] were enhanced. It has been observed that the increase in NaOH percentage and treatment time definitely gives high impact strength [38–41]. Most researches treated the bast fibers in a raw form, however there are different effects when treating the fibers alone or in woven fabric [42].

The effect of the alkaline treatment on the properties of the jute fabric is illustrated in Figure 4.2, that indicates the effect of the NaOH percentage concentration and the time of immersion of the fabric on its physical properties: all the measured properties are increased as the result of the NaOH treatment, reflecting on the mechanical properties of the jute fabric [42].

Fabric will have high crimp ratio, higher thickness, even the fabric areal density, and consequently the fabric density. The fabric morphology is completely changed with increased surface roughness caused by the increase in warp and weft crimp, as shown in Figure 4.3.

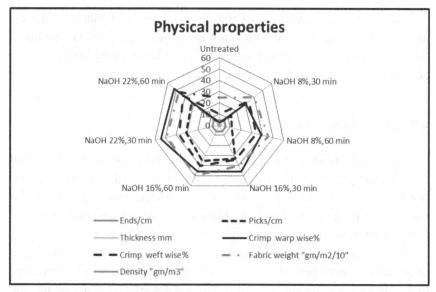

FIGURE 4.2   Physical properties of the differently treated fabrics.

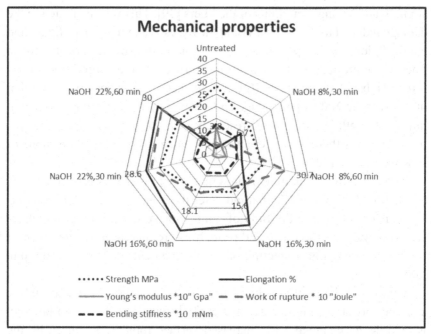

FIGURE 4.3 Mechanical properties of treated jute fabrics.

The load extension curve of the fabric shows that fabric's Young's modulus decreases while the breaking elongation increases as a result of fabric treatment, as illustrated in Figure 4.4. The breaking load is almost constant.

4.2.2.2 Coupling Agent

Coupling agent is defined as a compound which provides a chemical bond between two dissimilar materials, it usually improves the degree of cross-linking in the interface region and offers a perfect bonding. Among the various coupling agents, Silane coupling agents were found to be effective in modifying the natural fiber-matrix interface [43, 44].

SEM obtained for failure surfaces observed that with increasing in fiber–matrix interaction, the failure mode changes from interfacial failure to matrix failure. The mechanical properties determined from the tensile,

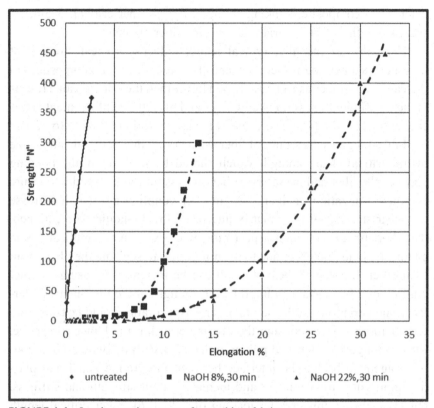

FIGURE 4.4   Load extension curve of treated jute fabrics.

flexural and shear tests exhibit a similar behavior for each of the different fiber surface treatments but the effect of the fiber surface treatment was more noticeable for the shear properties.

In the case of short fibers with length less than the critical fiber length given by the Eq. (1), the effect of the coupling agent will be higher than for the long fibers or threads and fabrics. If fiber length is greater than the critical fiber length, full utilization of the fiber strength contributes to the matrix overall strength. It is assumed that the fiber's length is completely bonded to the matrix, if not, the parts of the long fibers will debone and the long fiber will be considered as two or more fibers. Each may be of length less than the critical fiber length and the first condition will occur.

$$l_{critical} = \sigma_f d_f / 2\,\tau \tag{1}$$

where: σ_f – composite breaking strength, $l_{critical}$ – fiber critical length, τ – shear strength of fiber matrix interface, d_f – fiber diameter.

The value of the fiber critical length reduces as shear strength of fiber matrix interface τ_i increases and the strength of the composite will increases, too. Fiber aspect ratio (l_f/d_f) lower than the critical value results in the insufficient stress transfer to fiber and thus, the reinforcement is this case acts as a filler [44]. The value of the aspect ratio is $(l_f/d_f) : (\sigma_f/2\ \tau)$.

Table 4.4 gives the values of fiber aspect ratio, their strength, and value of the critical shear strength which should be reached by the bonding between the fiber and the matrix (fiber pulling out force) in order to insure complete utilization of fiber strength. The coupling agent will increase the shear strength of fiber matrix interface [45]. It should be mentioned, that short fibers with high aspect ratio will give some complications in composite manufacturing due to the difficulties to separate the fibers from each other. Interfacial shear strength can be increased by using coating/ grafting on fiber surface to improve the strength of the fiber-matrix interface. The variability of the natural fiber's aspect ratio should be taken into consideration when choosing the raw material for a polymer composite. Maleic-anhydride-modified polypropylene (MAPP) is found a good coupling agent to bridge the interface between the ground kenaf and plastic, improving stress transfer and increasing their strength and stiffness, but also allow a higher filler loading of 65% [46]. Fiber critical length

TABLE 4.4   Fiber Aspect Ratio and Critical Shear of Natural Fibers

Fiber	Fiber aspect ratio	Strength MPa	Critical shear N/cm²	Fiber	Fiber aspect ratio	Strength MPa	Critical shear N/cm²
Flax	1738	500–900	25.89	Pineapple	450	400–1600	177.78
Hemp	1000	300–800	40	Abaca	250	1100	220
Jute	100	200–500	250	Coir	1750	131–175	5
Kenaf	160	284–800	250	Cotton	900–1600	300–600	30
Ramie	1500	220–938	27.93	Soft wood	100	98–170	85
Sisal	20	100–800	2000	Hard wood	50	90–180	180
Banana	80	500–700	437.5				

and fiber aspect ratios obtained for the different fiber surface treatments, using the single fiber fragmentation test. For all the surface-treated fibers there was a noticeable decrease of the fiber critical length as compared to the untreated fibers. It was found that for the untreated fibers, an average critical length value of 12.96 mm, it can be reduced to 6 mm for the pre-impregnated fiber and 3.5 mm for the silane treated fiber [47]. One of the decisive factors that play role in improving the fiber matrix strength is the fiber surface area to weight ratio *(S/W)*. The surface area of the fibers "S" in cm for fibers of weight "w" in grams is:

$$(s/w) = 1121/ (tex. \rho)^{0.5} \tag{2}$$

where: ρ – fiber density g/cm^3, *tex* – fiber count.

The finer the fiber is, the higher will be value of fiber surface area per linear weight, resulting in the higher value of fiber-matrix interfacial strength, as illustrated in Figure 4.5.

At the present time, over 40 coupling agents have been used in the pre-treatment of natural fiber composites [47]. There are several effective methods for chemical modifications to improve the interfacial force between the fibers and matrix [4, 9, 39–41, 45–48], namely:

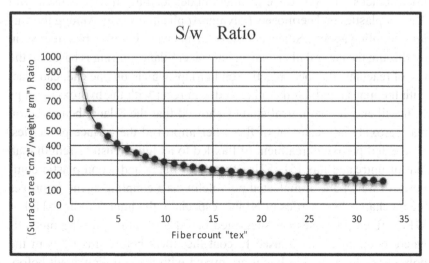

FIGURE 4.5   Ratio of surface area /weight of fiber versus fiber count.

1. Graft copolymerization. (Polypropylene-malice anhydride (MAH-PP) copolymers).
2. Treatment with compounds which contain methanol groups.
3. Treatment with isocyanates [polymethylene-polyphenyl-isocyanate (PMPPIC)].
4. Triazine coupling agents.
5. Silane as coupling agent.

In grafting copolymerization, after the treatment, the surface energy of the fibers is increased to a level much closer to the surface energy of the matrix. Maleic anhydride grafted polypropylene (MAPP) can be used in grafting copolymerization with a certain percentages, depending on the type of fiber, the optimum MAPP to fiber ratio is found, in average, to range between 10% and 13.3% to attain the best mechanical properties of the NFPC [49]. The difference with other chemical treatments is that maleic anhydride is not only used to modify fiber surface but also the PP matrix to achieve better interfacial bonding and mechanical properties in composites [17]. The PP chain permits maleic anhydride to be cohesive and produce MAPP, so that the treatment with cellulose improves the bonding with the matrix.

Silane, as a coupling agent, has the capability to build durable bonds between organic and inorganic materials to prevent the propagation of the micro cracks on the surface of fibers under loading. It can be used with Thermoplastics or Thermosets Polymer Applications [50]. Among the various coupling agents, Silane coupling agents were found to be effective in modifying the natural fiber-matrix interface. Silane grafting is based on the use of reactants that bear reactive end groups which, on one end, can react with the matrix and, on the other end, can react with the hydroxyl groups of the fiber. Morphological studies showed that the Silane, benzoylation and peroxide pretreatment of flax fiber improved the surface properties. Silane and peroxide treatment of flax led to a higher tensile strength than that of untreated flax [51, 52]. For natural fibers and wood plastic composites (WPC) 3% of maleated polyolefin, as a coupling agent, improves the mechanical properties of NFPC, especially the impact force [51]. For WPC mixed processes are suggested based on coating/grafting methods where coupling agent is used to coat the fibers before mixing with the polymer or fibers and polymer are mixed with the coupling agent before pre-treatment (coating or grafting). For instance, in WPC the coupling agent such as, MAPP-maleated polypropylene, PHA-phthalic anhydride, and Polymethacrylic acid (PMAA), represents 2–8%. The fiber volume fraction varies between 50–70%, depending on the fabrication method [54]. Mechanical properties of composite depend on the fiber matrix type (thermosets or thermoplastic) and type of coupling agent. The strength of the composite increases up to 30%. The impact resistance of the composite material increases up to 100% [53].

4.2.2.3   Grafting

4.2.2.3.1   *Esterification Process*

Etherification involves the reaction of an alcohol (here a saccharide alcohol) with an alkylating agent in the presence of a base [16]. The hydrophobization of the cellulose surface has usually been achieved through the well-known cellulose esterification process, which basically uses carboxylic acid, acid anhydrides or acyl chlorides as reacting agents. Esterification is a reaction that introduces an ester functional group ($O-C = O$) onto the surface of cellulose by condensation of the previous reagents with a cellulosic alcohol group. Acetylation is the reaction that introduces an acetyl functional group CH3-C (=O) – onto the surface of cellulose [54]. This basic reaction is also involved in the preparation of cellulose ester derivatives, such as the well-known cellulose acetate. Some investigators applied low-pressure plasma-processes to allow enhanced chemical vapor deposition [55].

4.2.2.3.2   *Polymer Grafting*

Chemical surface modification of cellulose fibers or nanoparticles can be achieved by covalently attaching small molecules, as well as polymers in order to increase a polar character of the fiber and have a better compatibility with hydrophobic polymer matrices [55, 56]. Two main approaches can be used to graft polymers onto surfaces; grafting onto, by addition of polymer chains onto a surface and used for Nanocrystalline Cellulose particles or grafting from, a polymer chain is initiated and propagated at the surface and used for Nanofibrillated Cellulose. Both methods grafting "onto" or "from" are the different ways to alter the chemical reactivity of the surface [56]. First approach involves mixing the cellulosic nanoparticles with an existing polymer and a coupling agent to attach the polymer to the nanoparticle surface. The second approach consists of mixing the cellulosic fibers or the activated cellulosic nanoparticles with a monomer and an initiator agent to induce polymerization of the monomer from the fiber surface.

4.3 NATURAL FIBER COMPOSITE MATRIX

4.3.1 POLYMERS FOR NATURAL FIBER COMPOSITE (NFPC) MATRIX

The polymer matrix is responsible to support the structure of the fiber reinforcement, giving the final neat shape of the designed part, aligning the fibers, transfer the load between the fibers, assisting the fibers in providing compression strength and modulus to the composites, assisting the fibers in providing shear strength and modulus, withstand heat, and protect the fiber reinforcement from the environmental conditions. The failure of the matrix to fulfill these major requirements will lead to failure of the composite under low stress. Besides that, it is preferable to be biodegradable and easily recycled. Figure 4.6 illustrates the stress strain curves of fiber and the polymer. High diversity in the properties of fiber and polymer reflects on the failure performance of the composite.

Other important property under consideration is the viscoelastic behavior of the polymer. The stress strain curve of the polymer is affected significantly by the temperature. At elevated temperature polymer mechanical properties will deteriorate due to the decrease of polymer molecular weight, thus decreasing the tensile strength, elastic modulus and increasing the ductility, impact strengths and viscosity. The change of these properties is a function of the molecular weight of the polymer. The elastic modulus decreases at higher rate after the polymer temperature increases over glass-transition temperature (temperature region where the polymer transitions from a hard, glassy material to a soft, rubbery material "T_g")

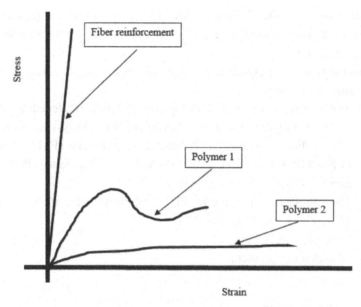

FIGURE 4.6   Stress–strain curve of fiber and polymer.

and dropped dramatically on reaching the melting temperature (T_m), the rate depends on the degree of crystallinity. Also, the presence of crosslinks decreases the dropping rate of the mechanical properties [57, 58].

4.3.2 CLASSIFICATION OF POLYMER TYPES FOR NFPC MATRIX

The construction of the NFPC aims to achieve the improved characteristics of fiber and polymer combination, such as strength, stiffness, shear resistance, creep resistance, hydrophobicity, dimension stability, solvent and acid resistance, fire retardant, surface abrasion, and biodegradability. Consequently, there are a certain polymer properties that are required according to end use of the composite, for instance, good bonding to the fibers, strength, flexibility, elastic recovery, low glass transition temperature, suitable melting temperature, easy to be adopted with the different techniques of composite manufacturing methods, weathering resistance. The glass transition temperature helps in the determination of the working temperature for the manufacturing process hence the properties of the polymer are quite different under the glass transition temperature and above it. The polymer's (T_g) is lower than its (T_m) approximately by 150°C, Table 4.5.

Several types of polymers are used for processing the matrix of the composites, such as:

1. Thermosets polymers; Polypropylene (PP), Polyethylene low density (LDPE), Polyethylene high density (HDPE), Poly(vinyl chloride) (PVC), Polystyrene (PS), Polytetrafluoroethylene (PTFE, Teflon), Poly(methyl methacrylate) (PMMA, Lucite, Plexiglas), Poly(vinyl acetate) (PVAc), Polychloroprene.
2. Thermoplastics polymers: PET Polyproplyene, Polycarbonate, PBT, Vinyl, Polyethylene, Nylon.
3. Biological polymers.
4. Elastomers polymers.

4.3.2.1 Thermosets

Thermosetting plastics are polymer materials which are liquid or malleable at low temperatures, but which change irreversibly to become hard at high temperatures hence, cure process requires to be induced by heat, generally above 200°C, depending on the type of polymer [60–61]. The degree of wetting during the production process is important for a good adhesion between fiber and matrix. The thermoset materials have glass transition temperature but no melting temperature. When applying thermosets the viscosity can be low T_g, this eases the wetting. For some lay-ups, the specific strength and stiffness will even be better compared to glass composite. Problems that can be encountered are related to moisture. The fiber moisture can affect the chemical reaction. In order to prevent this, the fibers have to be dried before, preferably down to 2–3%. In standard room condition, the moisture content is often over 10 percent. Air is always present in the fibers and in the resin. The surface of the natural fiber has geometry and a chemical condition on which air bubble growth will be initiated, especially in vacuum processes like vacuum injection. In order to prevent many voids and a poor fiber matrix interface during vacuum injection, it is necessary to dry the fibers and to degas the resin. Examples of thermosetting plastics are: Unsaturated polyester, Polyurethane Adhesives, Anaerobic Adhesives

TABLE 4.5 Some Physical Properties of Polymers [60–74]

Polymer	Density g/cm^3	Tensile strength (MPa)	Tensile elongation to break %	Young's modulus (GPa)	Melting temp., Tm (°C)	Glass transition temp. Tg (°C)	Poisson's ratio "v"
Polypropylene (PP)	0.9–0.94	27–33	200–700	1.3–4.4	134–174	–10	0.45
Low density Polyethylene (LDPE)	0.95	20–30	20–100	0.700	124–136	–125	0.4
High density Polyethylene (HDPE)	0.93–0.97	10–60	4–18	0.60–2.90	115–125	–80	0.40–0.46
Polystyrene (PS)	1.050	48	1–60	1.4–4	240	89–106	0.35
Polyvinyl chloride (PVC)	1.330	48	40–450	.014–4	212	80	
Acrylic	1.19	74	6	3	265	85 to 165	0.38
Nylon	1.15	55–83	60–200	1.4–2.8	265	47 – 60	0.32–0.40
General purpose epoxy	1.1–1.4	35–100	1–6	3–6	200–265		
Polyester	1.2–1.5	40–90	5–300	2–4.5	200–265	30	0.38

TABLE 4.5 (Continued)

Polymer	Density g/cm³	Tensile strength (MPa)	Tensile elongation to break %	Young's modulus (GPa)	Melting temp., Tm (°C)	Glass transition temp. Tg (°C)	Poisson's ratio "ν"
Polylactic acid (PLA) ***	1.21–1.43	58	4–8.7	3.5	145–186	63.8	0.36
Cellulose acetate ***	1.3	55–70	75–25	1.4–3.5	127–240	85–141	0.35–0.40
Starch plastics *&****	1.5	12.4	16	900		95**	0.419
Polycaprolactone (PCL) ***	1.146	1000	300	1200	60	60	0.3
Polyester amide ***	1.4	19.6	12	0.535	87.8	60	0.38–0.5

*Starch glitanzation temperature varied between 58–73°C and the pasting temperature 80–74°C [73].

**Depending on the moisture content.

***Biodegradable.

Araldite, Bakelite, Epoxy, Faturan, Melamine resin, Phenol formaldehyde resin, Polyester, Polyester resin, Polyhexahydrotriazine, Polyimide, Polyisocyanurate, Silicone, Urea-formaldehyde, Vinyl ester.

Thermoset polymers are usually harder, stronger, and more brittle than thermoplastics. They are insoluble in almost all organic solvents. The thermoset has high processing cost due to long processing, but easy in fiber impregnation, high resistance to creep, and resistance to heat and high temperature, fatigue strength, tailored elasticity, and excellent adhesion. While of the negative aspects of thermosets is it's no ability to recycle.

4.3.2.2 Thermoplastics

Thermosetting plastics are polymer materials solid at low temperatures, but which change to become soften on heating since secondly forces between the individual chains can break easily by heat or pressure. High temperatures can also cause unwanted changes of the polymer structure, most of the polymer is a mixture of amorphous and crystalline region the amorphous region will start to melt after the glass transition temperature Tg changing from leathery to liquid as temperature increased while the crystalline region will start to sharply melt after melting temperature Tm is reached.

Appropriate polymers or combinations of polymers have been investigated by several researchers to blend the polymer such as: Polyimide, Polycarbonate, Acrylic, Aramid, Polyester, and Orlon. The thermoplastic has high melt viscosity, difficult in fiber impregnation, high processing temperature and pressure but the manufacturing cost including mold cost for low volume production is low [61] and highly recyclable. Nevertheless, a low price, reasonable processing temperatures and recyclability are the reason for a growing interest in thermoplastic polymers, such as polypropylene. Unmodified PP however, will not have a proper adhesion with the fibers by applying consolidation forces alone. Mechanical properties are hardly improved: the fibers simply act like filler. Natural fibers will only act as reinforcement if compatibilizers are used. An interface between fiber and matrix should correct the natural rejection of both materials. An often used compatibilizer is MAPP, a modification of a PP chain with maleic anhydride. A small amount of MAPP added to the PP will lead to much higher strength properties of the material [59–61].

4.3.2.3 Biodegradable Polymers

The biodegradable composites can be divided into fully biodegradable and partly biodegradable [62] depending on the biodegradability of the matrix that capable of being decomposed by bacteria or other living organisms. Polylactide acid (PLA), thermoplastic starch, Cellulose esters, are examples of biodegradable polymer from natural source, while aliphatic polyester, aliphatic-aromatic polyester, polyvinyl alcohol, polyanhydrides, polyethylene terephthalate are examples of biodegradable polymer from petroleum based polymer. Partly degradable composites are those using non-degradable petroleum based polymer, such as polypropylene, polyester, polyethylene, polyvinyl alcohol. Biodegradable plastics and polymers were first introduced in 1980s. Polymers from renewable resources have attracted an increasing amount of attention hence Natural polymers are available in large quantities from renewable sources, while synthetic polymers are produced from nonrenewable petroleum resources. The natural Biodegradable polymers are directly extracted from the biomass or through organically modified organism [62]. In the last decade, the Biodegradable polymers, especially PLA, are widely used in manufacturing of NFPC and other industrial products [61–67].

4.3.2.3.1 Biodegradable Polymers Classification

The biodegradable polymers are classified according to their source: Natural resource or Synthetic [62–68], into categories depending on the synthesis and or the sources.

1. **The agro-polymers from agro-resources**:
   - Polysaccharides, such as starches, which is natural polymers are formed in nature during the growth cycles of all organisms. Natural biodegradable polymers are called biopolymers.

Polysaccharides, as starch and cellulose, represent the most characteristic family of these natural polymers. Wheat, potatoes, maize, lignocellulosic products, wood, straws, chitin, chitosan, gums, alginates.

- Poly peptide proteins, it may be prepared from animal sources, Collagen is the primary protein component of animal connective tissues. Several types of collagen exist, or from vegetal sources: Wheat gluten is a protein by-product of the starch fabrication, Soy protein and gluten. Chitin is usually found in the shells of crabs, shrimp, crawfish and insects. Collagen, gelatin, casein, and whey proteins. Collagen is enzymatically degradable and has unique biological properties.

2. **Bacterial polymers:** Polyesters obtained by polymerization of monomers prepared by fermentation process (semi-synthetic polymers) or produced by a range of microorganisms, Semi-synthetic polymers PLA, polyhydroxyalkanoates (PHA), Poly (hydroxybutyrate-co-hydroxyvalerate) (PHBV), Microbial polyesters, Poly-3-hydroxyalcanoates (PHB), Poly (Hydroxybutyrate-Hydroxyvalerate) (PHB/HV), Poly (Hydroxybutyrate), Poly-ε-Caprolactones (PCL).

3. **Polymers chemically synthesized:** using monomers obtained from agro-resources, such as PLA, Polyglycolic acid (PGA).

4. **Polymers** whose monomers and polymers are both obtained by chemical synthesis from petroleum resources, such as polycaprolactones (PCL), polyesteramides (PEA), aliphatic co-polyesters (PBSA), aromatic co-polyesters (PBAT), PGA, PLA and their copolymers, Polybutylene Succinate (PBS), Poly (Vinyl Alcohol) (PVOH) and Poly (Vinyl Acetate) (PVA).

Blends of Biodegradable Polymers Mixing biopolymers or biodegradable polymers with each other can improve their intrinsic properties. For example: Starch-based blends, Starch-poly (ethylene-co-vinyl alcohol), Starch-polyvinyl alcohol, Starch-PLA or to improve compatibilization is to use a compatibilizer like Maleic anhydride [69]. The compliance requirements for the key standards, such as ASTM is 60% of the material is biodegradable in period of 6 months. Testing for plastic materials according to ASTM D6400 or D6868 may require a minimum of 90 days, and a maximum of 180 days. Cording to Standard specification for compostable plastics.

4.3.2.4   Elastomers

Elastomers are the polymers with viscoelasticity and very weak intermolecular forces, generally having low Young's modulus and high failure strain compared with other materials. The elasticity is derived from the ability of the long chains to reconfigure themselves to distribute an applied stress [70]. Elastomers can extend from 5–700% with high elastic recovery. An example of the elastomer polymers are:

- Unsaturated rubbers: Natural polyisoprene, synthetic polyisoprene, polybutadiene, chloroprene rubber, polychloroprene, neoprene, and baypren.
- Saturated rubbers: Ethylene propylene rubber, ethylene propylene diene rubber, epichlorohydrin rubber, polyacrylic rubber, silicone rubber, fluorosilicone rubber, fluoroelastomers, tecnoflon, fluorel, perfluoroelastomers, polyether block amides, and ethylene-vinyl acetate.

Elastomer matrix may be used, like natural rubber or manmade rubber, in manufacturing of NFPC. This will combine the strength of the fibers and the impact properties of the rubber. It was found that rigid elastomer composite specimens had high impact resistance [71]. For sisal fiber reinforced rubber composite, reinforcement with short fibers offers some attractive features, such as high modulus and tear strength. Major factors which affect the performance of rubber-fiber composites are fiber loading, fiber dispersion, fiber orientation, fiber to matrix adhesion and the aspect ratio of the fiber [72]. Short fiber reinforced rubber composites were developed to fill the gap between the long fiber reinforced and particulate filled rubber composites. Composites in which the short fibers are oriented uniaxial in an elastomer have a good combination of good strength and stiffness from the fibers and elasticity from the rubber. These composites are being used for the fabrication of a wide variety of products such as v-belts, hoses and articles with complex shapes. Design flexibility is another advantage of these composites. Mechanical properties,

like specific strength and stiffness, reduced shrinkage in molded products, resistance to solvent swelling, abrasion, tear and creep resistance are
greatly improved in short fiber composites [73].

4.3.3 SELECTION OF POLYMERS FOR NFPC

The polymer properties that define the polymer can be expressed by chemical composition of repeat units, structural formula of repeat group, molecular weight of repeat unit, crystallinity, toxicity, environmental impact,
specific volume, molar volume, specific thermal expansion, thermal conductivity, melting temperature "tm," glass-transition temperature "Tg,"
cohesive properties, intrinsic viscosity, melt viscosity, diffusion coefficient, water absorption, hardness, bulk modulus, coefficient of friction,
tensile creep, elastic modulus, breaking elongation, fatigue, limits for fracture, flexural stiffness, flexural strength, hardness, impact strength, mold
shrinkage, Poisson's ratio, scratch resistance, shear strength, surface abrasion resistance, tear resistance, tensile strength break, Young's modulus,
toughness, ultimate strength, viscoelastic behavior, electrical resistance,
dielectric constant, biological stability, burning rate, flammability, chemical resistance, air and liquid permeability, thermal stability, UV resistance,
weathering. The rigid plastics, such as Polystyrene, Polymethyl methacrylate or Polycarbonate can withstand the applied stress on the composites,
but they can't withstand much elongation before breaking. Flexible plastics, like polyethylene and polypropylene, are different from rigid plastics.
They have low strength with high initial modulus, thus they will resist
deformation. In the case of a composite material, a fiber to reinforce a
thermoset is usually used. The fiber increases the tensile strength of the
composite, while the matrix gives compressional strength and toughness.

Some polymer properties are essential for the compatibility with the
end use to fulfill matrix performance under various loading conditions,
the manufacturing processes requirements, and life cycle analysis of the
product. Some properties of polymers used for composite are given in
Table 4.5.

The glass transition temperature (T_g) is an important property for designers to be aware of, since it lowers the expectations for mechanical performance at elevated temperatures. The high glass-transition temperature

has two contradictory effects – from one side the polymer will have good mechanical properties, on the other hand the cooling rate during curing should be slow. For NFPC the polymer melting temperature recommended not be high to protect the properties of the fiber.

The polymer properties will affect its mechanical properties and process ability. For instance, the increase of the polymer density will lead to increase the strength, stiffness, abrasion resistance, hardness, chemical resistance and, in the meantime, decrease of impact strength and crack resistance. While the increase of melt flow rate of the polymer decreases the most of the mechanical properties. But the reflection of these two properties on the manufacturing of the NFPC will be in different aspect hence the increase of the density of the polymer requires the increase of processing temperature and pressure, on the other hand the low melting flow rate improves the polymer impregnation through the reinforcement.

The following are some types of polymers used when targeting a specific composite matrix property, such as:

I. **For mechanical strength:** Nylon, Polyester, Polypropylene, Epoxies

II. **For stiffness:** Polyester, Polypropylene, Polystyrene

III. **For electrical resistance:** Nylon, Polycarbonate, Polyester, Polypropylenes, Rubbers

TABLE 4.6 Suitable Polymers for NFPC

Matrix	Fiber	Matrix	Fiber
Epoxy	Abaca, Bamboo, Jute	Polypropylene	Flax, Jute, Kenaf, Hemp, Wheat Straw, Wood fiber
Natural Rubber	Coir, Sisal	Polystyrene, Polyurethane, Polyvinyl chloride	Wood fibers
Nitrile Rubber, Phenol-formaldehyde	Jute	Polyester	Banana, Jute, Pineapple, Hemp
Polyethylene	Kenaf, Pineapple, Sisal, Wood fiber	Styrene-butadiene	Jute
Rubber	Oil palm	PLA	Wood fibers

IV. For heat resistance: Epoxies, Polyimides

V. For chemical resistance: Nylon, Polycarbonate, Polyester, Polypropylene.

From the experience of several researches [57, 74–78], the suggested polymers for different types of natural fibers are given in Table 4.6.

From Table 4.6, it clear that Polypropylene matrix is most popular for manufacturing of Natural fiber polymer composites.

KEYWORDS

- **biodegradable polymers**
- **coupling agent**
- **elastomers**
- **grafting**
- **plasma treatment**
- **selection of polymers**
- **surface treatment**
- **thermoplastics**
- **thermosets**

REFERENCES

1. Cicala, C.; Cristaldi, G.; Recca, G.; Latteri. A. Composites based on natural fiber fabrics, Woven Fabric. Engineering Polona Dobnik Dubrovski (Ed.), INTECH [Online] 2010 http://www.intechopen.com/books/woven-fabric-engineering/composites-based-on-natural-fiber-fabrics (accessed May 7, 2016).

2. Adekunle, K. F. Surface Treatments of Natural Fibers-A Review: Part 1. J. Polym. Chem. [Online] 2015, 5, 41–46 http://www.scirp.org/journal/PaperInformation.aspx?paperID=58642 (accessed May 7, 2016).

3. Anselm, O.; Ogah, N.; Nduji, A. A. Characterization and Comparison of Rheological Properties of Agro Fiber Filled High-Density Polyethylene Bio-Composites, J. Polym. Chem. [Online] 2014, 4, 12–19 http://www.scirp.org/journal/PaperInformation.aspx?PaperID=43380 (accessed May 7, 2016).

4. Mishra, S.; Misra, M.; Tripathy, S. S.; Nayak, S. K.; Mohanty, A. K. Potentiality of pineapple leaf fiber as reinforcement in Palf-polyester composite: surface modification and mechanical performance. J. Reinf. Plast. Compos. 2001, 20(4), 321–334.

5. Ansari, M. N. M.; Pua, G.; Jawaid, M; Islam, M. S. Review on natural fiber reinforced polymer composite and its applications. Int. J. Polym. Sci. [Online] 2015, Vol. 2015, 1–15 http://www.hindawi.com/journals/ijps/2015/243947/ (accessed May 7, 2016).

6. Bismarck, A.; Aranberri-Askargorta, I.; Springer, J. Surface Characterization of Flax, Hemp and Cellulose Fibers: Surface Properties and the Water Uptake Behavior. Polym. Compos. 2002, 23(5), 872–894.

7. Pai, A. R.; Jagtap, R. N. Surface Morphology & Mechanical properties of some unique Natural Fiber Reinforced Polymer Composites- A Review. J. Mater. Environ. Sci. [Online] 2015, 6(4), 902–917. http://www.jmaterenvironsci.com/Document/vol6/vol6_N4/106-JMES-869–2014-Pai.pdf (accessed May 7, 2016).

8. Gassan J, Bledzki A. K. Influence of fiber surface treatment on the creep behavior of jute fiber-reinforced polypropylene. J. Thermoplast. Compos. Mater. 1999; 12(5), 388–398.

9. Mukhopadhyay, S.; Fangueiro, R. Physical modification of natural fibers and thermoplastic films for composites – A Review. J. Thermoplast. Compos. Mater. 2009, 22(2), 135–162.

10. Bledzki, A.; Reihmane, K.; Gassan, J. J. Properties and modification methods for vegetable fibers for natural fiber composites. J. Appl. Polym. Sci. 1996, 59(8), 1329–1336.

11. Bledzki, A. K.; Gassan, J. Composites reinforced with cellulose based fibers. Prog. Polym. Sci. 1999, 24, 221–274.

12. Loan, D. T. T. Investigation on jute fibers and their composites based on polypropylene and epoxy matrices. PhD Theses, Technischen Universität Dresden, 2006.

13. Sparavigna, A. Plasma treatment advantages for textiles. [Online] http://arxiv.org/ftp/arxiv/papers/0801/0801.3727.pdf (accessed May 7, 2016).

14. Van den Oever, I. Plasma treatment of flax and hemp fibers to improve adhesion with polymers for composite applications. [Online] https://www.wageningenur.nl/en/show/Plasma-treatment-of-flax-and-hemp-fibers-to-improve-adhesion-with-polymers-for-composite-applications.htm (accessed May 7, 2016).

15. Sinha, E.; Panigrahi, S. Effect of Plasma Treatment on Structure, Wettability of Jute Fiber and Flexural Strength of its Composite. J. Compos. Mater. 2009, 43(17), 1791–1802.

16. Cumpstey, I. Chemical Modification of Polysaccharides, ISRN Organic Chemistry. [Online] 2013, 1–27. http://www.hindawi.com/journals/isrn/2013/417672/ (accessed March 10, 2015).

17. Li X.; Tabil, L. G.; Panigrahi, S. Chemical Treatments of Natural Fiber for use in natural fiber-reinforced composites: A Review. J. Polym. Environ. 2007, 15, 25–33.

18. Jazbec, K.; Šala, M.; Mozetič, M.; Vesel, A.; Gorjanc, A. Functionalization of cellulose fibers with oxygen plasma and ZnO nanoparticles for achieving UV protective properties. J. Nanomater. [Online] 2015, 1–9. http://www.hindawi.com/journals/jnm/2015/346739/ (accessed May 7, 2016).

19. Bruna Barra, B.; Paulo, B.; Alves, C. Effects of Methane Cold Plasma in Sisal Fibers, Key Eng. Mater. [Online] 2012, 517, 458–468. http://www.usp.br/constrambi/

producao_arquivos/20130416_atualizacao/Effects%20of%20Methane%20Cold%20 Plasma%20in%20Sisal%20Fibers.pdf (accessed May 7, 2016).

20. Fadel, N.; Aloufi, A.; El Messiry, M.; Militký, J. Atmospheric Plasma Treatment Stimulus of Wettability of Compact and Ring Spun Yarn. 8th International Conference – TEXSCI 2013, Liberec, Czech Republic, 2013.

21. Oraji, R. The effect of plasma treatment on flax fibers, MSc. Theses, University of Saskatchewan, 2008.

22. Aparecido, P.; Giriolli, J.; Amarasekra, J.; Moraes, G. Natural fibers plastic composites for automotive applications. [Online] http://www.speautomotive.com/SPEA_ CD/SPEA2008/pdf/c/BNF-04.pdf (accessed May 1, 2016).

23. Kumar, R.; Obrai, S.; Sharma, A. Chemical modifications of natural fiber for composite material. Pelagia Research, Der Chemica Sinica. [Online] 2011, 2(4), 219–228. http://pelagiaresearchlibrary.com/der-chemica-sinica/vol2-iss4/DCS-2011-2-4-219-228.pdf (accessed May 1, 2016).

24. Petinakis, E.; Yu, L.; Simon, G.; Dean, K. Natural fiber bio-composites incorporating poly (lactic acid) fiber reinforced polymers – The technology applied for concrete repair. Edited by Martin Alberto Masuelli, Publisher: InTech. [Online] 2013. http:// www.intechopen.com/books/fiber-reinforced-polymers-the-technology-applied-for-concrete-repair (accessed May 1, 2016).

25. Chavan, K. Chemical modifications of natural fibers for composite applications. [Online], http://www.slideshare.net/ketkichavan/chemical-modifications-of-natural-fibers-for-composite-applications (accessed May 1, 2016).

26. Bongarde, U. S.; Shinde, V. D. Review on natural fiber reinforcement polymer composites. IJESIT [Online] 2014, 3(2), 431–435. http://www.ijesit.com/Volume%203/ Issue%202/IJESIT201402_54.pdf (accessed May 1, 2016).

27. Ahad, N.; Parimin, N.; Mahmed, N.; Ibrahim, S.; Nizzam, K.; Mon Ho, Y. Effect of chemical treatment on the surface of natural fiber. Journal of Nuclear and Related Technologies, Special Edition, 2009, 6(1), 155–158.

28. Oladele, I. O.; Omotoyinbo, J. A.; Adewara, J. O. T. Investigating the effect of chemical treatment on the constituents and tensile properties of sisal fiber. J. Min. Mater. Eng. [Online] 2010, 9(6), 569–582. http://file.scirp.org/pdf/ JMMCE20100600007_11354020.pdf (accessed May 1, 2016).

29. Machaka, M.; Basha, H.; Chakra, H.; Elkordi, A. Alkali Treatment of Fan Palm Natural Fibers For Use In Fiber Reinforced Concrete. Eur. Sci. J. [Online] 2014, 10(12), http://eujournal.org/index.php/esj/article/view/3198 (accessed May 1, 2016).

30. Hashim, M.; Roslan, M.; Amin, A.; Zaidi A.; Ariffin, S. Mercerization Treatment Parameter Effect on Natural Fiber Reinforced Polymer Matrix Composite: A Brief Review. World A. Sci. Eng. Technol. [Online] 2012, 6(8), 778–784. http://waset.org/ publications/4608/mercerization-treatment-parameter-effect-on-natural-fiber-reinforced-polymer-matrix-composite-a-brief-review (accessed May 1, 2016).

31. Razak, N. I. A.; Ibrahim, N. A.; Zainuddin, N.; Rayung, M.; Saad, W. Z. The Influence of chemical surface modification of kenaf fiber using hydrogen peroxide on the mechanical properties of biodegradable kenaf fiber/poly (Lactic Acid) composites. Molecules 2014, 19, 2957–2968.

32. Xue, L.; Tabil, L. G.; Panigrahi, S. Chemical treatments of natural fiber for use in natural fiber reinforced composites: A Review. J. Polym. Environ. 2007, 15, 25–33.

33. Anike, D. C.; Onuegbu, T. U.; Ogbu, I. M.; Alaekwe, I. O. The effect of alkali treat-
    ment on the tensile behavior and hardness of raffia palm fiber reinforced composites.
    Am. J. Polym. Sci. 2014, 4(4), 117–121.
34. Zimniewska, M.; Przybylak-Wladyka, M.; Mankowski, J. Improvement of quality
    of bast fiber for diversified utilization. [Online] www.researchgate.net/…Bast (ac-
    cessed May 1, 2016).
35. Zhou L.; Yeung K.; Yuen C. Effect of NaOH mercerization on the crosslinking of
    ramie yarn using 1, 2, 3, 4-butanetetracarboxylic acid. Text. Res. J. 2002, Vol. 72, No.
    6, 531–538.
36. Modibbo U.; Aliyu B.; Nkafamiya I. The effect of mercerization media on the physical
    properties of local plant bast fibers, International Journal of Physical Sciences. [Online]
    2009, 4(11), 698–704. http://www.academicjournals.org/ijps (accessed May 1, 2016).
37. Wang, W.; Cai, Z.; Yu, J. Study on the chemical modification process of jute fiber.
    JEFF 2008, 3(2), 1–11.
38. Hai N. M.; Kim B.; Lee S.; Effect of NaOH treatments on jute and coir fiber pp com-
    posites. Adv. Compos. Mater. 2009, 18(3), 197–208.
39. Rodríguez E. S.; Vazquez A.; Alkali treatment of jute fabrics: Influence on the pro-
    cessing conditions and the mechanical properties of their composites. The 8th Inter-
    national Conference on Flow Processes in Composite Materials (FPCM8), Douai,
    France 11–13 July 2006, 105–112.
40. Sutharson, B.; Rajendran, M.; Devadasan, S.; Selvam B. Effect of chemical treat-
    ments on mechanical properties of jute fiber hybrid composite laminates. ARPN
    Journal of Engineering and Applied Sciences. [Online] 2012, 7(6), 760–765. http://
    www.arpnjournals.com/ (accessed May 4, 2016).
41. Nam G., Kim J., Byeon J., Kim B., Kim T., Song J., Effect of surface treatment on
    mechanical behavior of jute fiber-reinforced polypropylene composite. 18th Inter-
    national conference on composite materials ICC18 (Jeju Island) Korea, The Korean
    Society of Composite Materials, 2011.
42. El Messiry, M.; Fadel, N. Modification of the jute fabrics for light weight cemen-
    tations composites (TRC). Aachen-Dresden International conference, Aachen, Nov.
    29–30, 2015, Aachen.
43. Xiea, Y.; Hillb C. A. S.; Xiaoa, Z.; Militza, H.; Maia, C. Silane coupling agents
    used for natural fiber/polymer composites: a review, Composites Part A 2010, 41(7),
    806–819.
44. Botros, M. Equistar Chemicals, LP, Development of new generation coupling
    agents for wood-plastic composites. [Online] https://www.researchgate.net/publi-
    cation/237677324_development_of_new_generation_coupling_agents_for_wood-
    plastic_composites (accessed May 4, 2016).
45. PHYS.ORG. Kenaf powder used to create new durable wood-plastic composite ma-
    terial. [Online] http://phys.org/news/2012–10-kenaf-powder-durable-wood-plastic-
    composite.html (accessed May 4, 2016).
46. Ichhaporia, P. K. Composites from natural fibers. PhD Theses, Fiber and Polymer
    Science Dept. NCSU, 2008.
47. Herreren-Franco, P. J.; Valadez-Conzalaz, A. A study of the mechanical properties of
    short natural-fiber reinforced composite. Composite B 2005, 36, 597–608.

48. El-Sabbagh, A. Effect of coupling agent on natural fiber in natural fiber/polypropylene composites on mechanical and thermal behavior. Composites Part B: Engineering, 2014, 57, 126–135.
49. Gelest Inc, http://www.gelest.com/goods/pdf/couplingagents.pdf (accessed May 1, 2016).
50. Cicala, G.; Cristaldi, G.; Recca, G.; Latteri. A. Composites based on natural fiber fabrics, woven fabric engineering. Dubrovski, P. D. (Ed.), 2010.
51. Xiea, Y.; Hill, C. A. S.; Xiaoa, Z.; Militza, H.; Maia, C. Silane coupling agents used for natural fiber/polymer composites: a review. Composites A: Applied Science Manufacturing, Composites: Part A 2010, 41, 806–819.
52. Eastman chemical Co. http://www.eastman.com/Literature_Center/A/APG2.pdf (accessed May 2, 2016).
53. Ziqiang, L. Chemical Coupling in Wood-Polymer Composites, PhD Theses, Louisiana State University, 2003.
54. Hermans, J. J. Chemical Mechanisms in the Grafting of Cellulose. Pure and Applied Chemistry 1962, 5(1–2), 147–164.
55. Favia, P.; D'Agostino; F. Palumbo. Grafting of Chemical Groups onto Polymers by Means of RF Plasma Treatments: a Technology for Biomedical Applications. Journal de Physique 1997, IV, 07 (C4), 199–208.
56. Missoum, K.; Belgacem, M. N.; Bras, J. Nano fibrillated Cellulose Surface Modification: A Review, Materials. [Online] 2013, 6, 1745–1766. http://www.mdpi.com/1996–1944/6/5/1745/htm (accessed May 5, 2016).
57. Nicholson, L. M.; Whitley, K. S.; Gates, T. G.; Hinkley, J. A. How molecular structure affects mechanical properties of an advanced polymer [Online] http://ntrs.nasa.gov/archive/nasa/casi.ntrs.nasa.gov/20040086993.pdf (accessed May 5, 2016).
58. Mohammed, L.; Ansari, M. N. M.; Pua, G.; Jawaid, M.; Islam, M. S. A Review on Natural Fiber Reinforced Polymer Composite and Its Applications, Int. J. Polym. Sci. [Online] 2015, 1–15. http://www.hindawi.com/journals/ijps/2015/243947/ (accessed May 5, 2016).
59. Hamour, N.; Boukerrou, A.; Djidjelli, H.; Maigret, J.; Beaugrand, J. Effects of MAPP Compatibilization and Acetylation Treatment Followed by Hydrothermal Aging on Polypropylene Alfa Fiber Composites. Int. J. Polym. Sci. [Online] 2015, 1–9. http://www.hindawi.com/journals/ijps/2015/451691/ (accessed May 2, 2016).
60. Brouwer, W. D. Natural fiber composites in structural components: alternative applications for sisal? FAO, Economic and Social Development Department http://www.fao.org/docrep/004/y1873e/y1873e0a.htm#TopOfPage (accessed May 2, 2016).
61. Masuelli, M. Introduction of fiber-reinforced polymers − polymers and composites: concepts, properties and processes, fiber reinforced polymers − the technology applied for concrete repair, Ed. Masuelli, M. A., Pub. InTech, 2013.
62. Ghanbarzadeh. B.; Almasi, H. Biodegradable Polymers, Biodegradation − Life of Science, Chapter 6, Ed. Chamy, R., Pub. InTech. [Online] 2013. http://www.intechopen.com/books/biodegradation-life-of-science/biodegradable-polymers (accessed May 5. 2016).
63. Jacobson, S.; Fritz, H. G. Plasticizing Polylactide − the effect of different plasticizers on the mechanical properties. Polym. Eng. Sci. 1999, 39, 1303–1310.

64. Avérous, L.; Pollet, E. Bio composite Biodegradable Polymers, chapter 2. Springer pub.2012.
65. Ranucci, E. Polymer chemistry and environmentally degradable polymers. Org. biochem. Vol. II: polymer chemistry and environmentally degradable polymer. [Online] http://www.eolss.net/sample-chapters/c06/e6-101-12-00.pdf (accessed May 6, 2016).
66. Vroman, I.; Tighzert, L. Biodegradable Polymers. Mater. [Online] 2009, 2, 307–344 file:///C:/Users/magdy/Downloads/materials-02-00307%20(5).pdf (accessed May 5, 2016).
67. Rathnam, K. V.; Peel, L. D. Impact resistant fiber reinforced elastomer composite materials, Proceedings SAMPLE 2004, Long Beach CA. USA, 2004.
68. Vieira, M. G. A.; da Silva, M. A.; Santos, L. O.; Beppu, M. M. Natural-based plasticizers and biopolymer films: A review. Eur. Polym. J. 2011, 47(3), 254–263.
69. Joseph, K.; Filho, R. D. T.; James, B.; Thomas, S.; de Carvalho, L. H. A. Review On Sisal Fiber Reinforced Polymer Composites, Revista Brasileira de Engenharia Agrícola e Ambiental 1999, 3(3), 367–379.
70. Elastomer https://en.wikipedia.org/w/index.php?title=Elastomer&oldid=712634376 (accessed May 5, 2016).
71. Jacobson, S.; Fritz, H. G. Plasticizing Polylactide-The Effect of Different Plasticizers on the Mechanical Properties. Polym. Eng. Sci. 1999, 39(7), 1303–1310.
72. Schmid, K. Manufacturing Processes for Engineering Materials, 5th ed. Pearson Education 2008. http://www3.nd.edu/~manufact/MPEM_pdf_files/Ch10.pdf (accessed May 1, 2016).
73. Mathew, L. Short Isora fiber reinforced natural rubber composites. PhD Theses Department of Polymer Science and Rubber Technology Cochin University of Science and Technology Kochi 22, Kerala, India, 2006.
74. Thongsane, P. The properties Improvement of rice starch by cooperated with cellulose and crystalline cellulose from palm pressed fibers. PhD Theses, Prince of Songkla University, Thailand, 2009.
75. Ku, H.; Wang, H.; Pattarachaiyakoop, N.; Trada, M. A review on the tensile properties of natural fiber reinforced polymer composites. Composites Part B 2011, 42, 856–873.
76. Van Vuure, A. Recent Development in Natural Fiber. [Online] http://www.i-sup2008.org/presentations/Conference_1/VanVuure_AW.pdf (accessed May 6, 2016).
77. Malkapuram, R; Kumar, V.; Neg, Y. Recent Development in Natural Fiber Reinforced Polypropylene Composites. Compos. 2009, 28(10), 1169–1189.
78. Gopinath, A.; Kumar. M. S.; Elayaperuma A. Experimental Investigations on Mechanical Properties of Jute Fiber Reinforced Composites with Polyester and Epoxy Resin Matrices. Procedia Eng. 2014, 97, 2052–2063.

CHAPTER 5

SOME ASPECTS OF TEXTILE COMPOSITE DESIGN

CONTENTS

5.1   INTRODUCTION

The structure of the composite is very simple – it consists of three-dimensional preforms that form the skeleton of the designed part. The reinforcement can be in form of fiber mat, individual fibers, yarn, weave structure, woven fabric, knitted fabric, braided fabric, nonwoven fabric, and triaxial fabric. Also, it may be 2-D or 3-D fabrics. The matrix binds the fibers, holding them in the pre-determined form, to protect the reinforcement from the environment effect, as well as to transfer the applied loads in all the directions to all the fibers in the preform as the reinforcement is the main bearing load element. The interface between the matrix and the fiber plays the important role in the transfer the loads from the matrix to the fibers. Hybrid composites, which may be formed from several reinforcements of different material or filler [1, 2], are presented in single matrix, as shown in Figure 5.1. The variety of the components for a composite are usually used to reach the properties which can't be obtained with single reinforcement or single matrix.

The fiber polymer composite have the following advantages: high specific strength, high specific modulus of elasticity, fatigue resistance, creep resistance, low coefficient of thermal resistance, corrosion resistance, design flexibility, low cost, less noisy while in operation, versatile than metals in complex designs, high corrosion resistance and ecofriendly. A new application of textile – steel reinforced composites have been introduced in some automotive applications [3]. The textile composites (FPC) can be classified as:

- Synthetic fiber composite, synthetic fiber reinforcement
- Natural fiber composite, natural fiber reinforcement
- Bio-composite, both reinforcement and matrix are environment-friendly biodegradable materials.

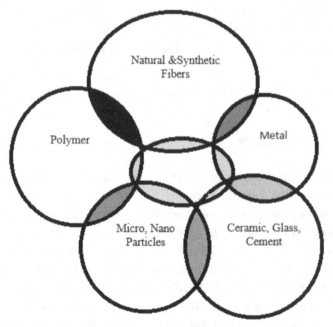

FIGURE 5.1   Various composite structure combinations.

5.1.1   DESIGN PROCEDURES OF COMPOSITE

According to the end use of the composite, the designer will determine the forces acting on the designed part and start to follow the steps given in Table 5.1, which describes the procedure for designing of textile composite. Determining the material requirements for certain component of product in the design should be based on the specifications and end use [4].

Each of the requirements of final composite is a function of both the fiber and matrix properties as well as the pretreatment methods.

5.1.2   SELECTION OF FIBERS

The first step in the design of the composite and the performance of the designed part will depend on the choice of the materials to withstand:

TABLE 5.1   Procedures to Design Textile Composite

Steps	Parameters
1. Selection of fiber properties	Physical properties, Mechanical properties, Thermal properties
2. Selection of reinforcement	Material: Flax, jute, hemp, rami, kenaf, soft wood
	Architect: Particle, whiskers, fiber, yarns, woven fabric, knitted fabric, nonwoven fabric
3. Selection of pretreatment of reinforcement	Cold Plasma Treatment, Alkaline Treatment, Esterification, Salinization, Acid Treatment, Mechanical treatment
4. Selection of coupling agent	Vinyl, chloropoeopl, methacryl, amine, cationic, phenol, mercpto, phosphate, neoaloxy, silane, Maleic anhydride
5. Selection of matrix material	Thermoplastic (poly ethaline, polypropylene, polyamide), thermoset (epoxy, vinyl ester, phenolic) and bio-gradable (natural source resin) composite
6. Selection of manufacturing technology and parameters	Hand layup, spraying, compression, transfer, resin transfer, injection, compression injection, pressure bag molding, centrifugal casting, cold press molding, continuous laminating, filament winding, pultrusion, extrusion
7. Composite testing	Type of tests according to the loading conditions

1. Type of applied forces;
2. Condition of applied forces;
3. Working environment;
4. The joint methods to the other parts;
5. Method of manufacturing;
6. Serviceability;
7. Cost; and
8. Environmental issues.

The analysis of the different factors which reflect the effect of material properties on the product specifications are analyzed, Table 5.2.

The mechanical properties required for the design of composite of most common natural fibers used for NFPC are given in Table 5.3.

The density, specific strength and specific Young's modulus of the natural fibers in comparison to glass fibers, illustrated by Figure 5.2, indicate that the specific Young's modulus of most natural fibers has almost the same value, while the specific strength of flax, hemp, jute and bamboo fibers have highest values. Flax fiber can be a replacement to the E-glass fiber in some applications.

TABLE 5.2 Material Selection for Composite Design

Composite requirements	Fiber properties			
Performance	Fiber strength and elongation	Fiber stiffness	Fiber impact strength	Fiber and matrix creep properties
Weight	Fiber density	Dimensions		
Recycling	CO_2 emission	Possibility of reuse	Safety of disposal	
Bio-degradation	Effect of the environment on the degradation of the fiber and matrix	Rate of degradation		
Environmental issue	Energy consumption in composite elements manufacturing and forming	CO_2 emission		
Cost	Raw material cost	Manufacturing cost	Recycling cost	

TABLE 5.3 Mechanical Properties of Natural Fibers as Compared to Conventional Reinforcing Glass Fiber

Properties	Flax	Hemp	Jute	Ramie	Coir	Sisal	Kenaf	Cotton	Soft wood	Hard wood	Bamboo	E-glass
Density g/cm³ "ρ"	1.4	1.48	1.46	1.5	1.25	1.33	1.193	1.5	1.5	1.2	0.8	2.55
Tensile strength MPa	600–1500	550–900	400–800	500	220	600–700	240–600	400	1000		465	2400
Young's modulus												
"E" GPa	60–80	70	10–30	44	6	38	14–38	12	40	37.9	18–55	73
Specific modulus (E/ρ)	26–46	47	7–21	29	5	29	23	8	26.7	31.6	22.5–68.8	29
Breaking Elongation %	1.2–1.6	1.6	1.8	2	15–25	2–3	1.6	3–10			2.5–3.7	1.4

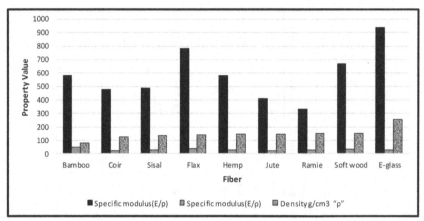

FIGURE 5.2   Comparison of the natural fibers properties with E-glass properties.

5.1.3 SELECTION OF FABRIC

Continuous fiber reinforced composites are now firmly established engineering materials for the manufacture of components in the automotive and aerospace industries [4]. In this respect, composite fabrics provide flexibility in the design manufacture. Many investigations point out that textile fabric possesses hardly any stiffness in bending or compression, so it is able to cover a 3-D body gracefully and can deform to a complex shape easily. However, because of its small bending rigidity, it cannot support compressive stresses. When sheet of textile material deforms in a compressive direction, buckling occurs and wrinkles are formed. Wrinkling (i.e., buckling of fibers), arising during the forming of textile composites, tends to significantly degrade the performance characteristics of the final product.

The 3-D fabric's inherent three-dimensional structure is creating integrity in three mutually perpendicular directions [5]. 3D-fabric reinforcement can overcome the problem of low inter-laminar and through-thickness strength typical for traditional 2D-laminates. The mechanical properties of composites made from fiber laminate are characterized by high in-plane stiffness and strength and lower out-of-plane stiffness and strength [6]. One of the main advantages of 3D-fabric reinforcements is the increased fracture toughness. The fracture toughness is a measure of how well a material containing cracks can resist fracture. The increased resistance to delamination also has a positive effect on compression strength after impact [7]. The deformation modes of composite fabrics during the forming process are different than those of sheet metal. Number of deformation mechanisms are available, including shear deformation between warp and weft fibers, fiber straightening, relative fibers slip, and yarn buckling [4]. Any type of fabric can be used, with the appropriate manufacturing method, when it possesses the following properties [7]:

- High compression resistance;
- Low fabric bending resistance;
- Anisotropy in mechanical properties;
- Low between yarn coefficient of friction;
- Low shear modulus;
- High value of bursting strength;
- High value of melting temperature;
- High value of tensile properties;
- Suitable porosity;
- High thermal resistance.

For knitted fabric, shear modulus is much lower than for the woven or braided fabrics due to its loose structure and the difference of the mechanical properties in the direction of walls and courses. This makes it more suitable for the use in 3-D preform formations. Moreover, the use of Lycra yarn in the knitted fabric gives additional decrease of the shear modulus. Thus, the most suitable type of fabric for the manufacturing of 3-D complicated shapes for composite application are the knitted fabrics [8].

5.1.4 SELECTION OF COMPOSITE STRUCTURE

The composite is formed of the laminates from several layers of woven fabric. The direction of warps in each laminate may be unidirectional or take different angles in each laminate, as shown in Figure 5.3. Usually, the direction of laminate is the direction of laminate's higher strength, for example in woven fabric, the direction of warp. Hence, layout of laminate in the composite, Figure 5.3, can be expressed by the angle of inclination

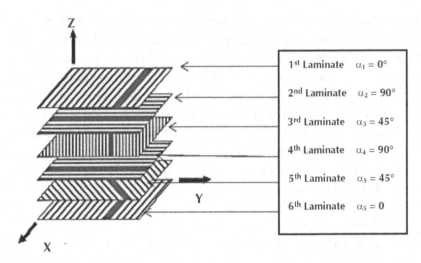

FIGURE 5.3   Laminate architect of the composite.

of the warp yarns to the x-axis angle "α_i," where "i" is the order of the laminate in the composite.

The discretion of such layer arrangement can be expressed as: $[0°, 90°, 45°, 0°, 45°, 0°]$, which is called "composite laminate code." Symmetrical laminate lay, $[0°, 45°, -45°, 90°, 90°, -45°, 45°, 0°]$, which is symmetrical in construction relative to its neutral axis, thus it can be written as: $[0°, 45°, -45°, 90°]$.

The code of the composite will indicate the percentage of the laminate that has the same direction of the warp yarns. This will affect the load bearing of the composite in each direction. The shear force at the surface of each laminate depends on laminate code, too. This will impact on the delamination failure of the composite [9]. Laminate is called quasi-isotropic when its extensional stiffness matrix behaves like an isotropic material.

The angle between the fiber orientations is given as:

$$\Delta\Theta = \pi / N$$

where: N – the number of laminates.

The quasi-isotropic laminate with the construction for $N = 3, 4, 6,$ and 12 will have fiber orientations $\Delta\Theta = 60°$, $\Delta\Theta = 45°$, $\Delta\Theta = 30°$, and $\Delta\Theta = 15°$.

The delamination failure of textile composite will be more noticeable regarding Hybrid composites, where several reinforcements of different materials are presented in single matrix or when one reinforcement is used with a mixture of different materials [9].

5.1.5 THE CHOICE OF MATRIX POLYMER

The selection of the type of matrix has a profound influence on the mechanical properties of the composite. A designer is always interested in the estimation of failure stresses of the material he wants to employ in his design. The most important characteristics, requiring consideration for most engineering components, include mechanical properties (strength, stiffness, specific strength, stiffness, fatigue, toughness, and the influence of high or low temperatures on these properties) and degradation. Special properties, for example, chemical, thermal, electrical, optical and magnetic properties, damping capacity, etc., methods of fabrication and the total costs attribute to the material selection and manufacturing technique.

5.2 STRESS ANALYSIS IN COMPOSITE

5.2.1 GENERAL STRESS IN MATERIAL

The stress on orthogonal surfaces, assuming composite material is a homogenous material, when subjected to forces is as shown in Figure 5.4.

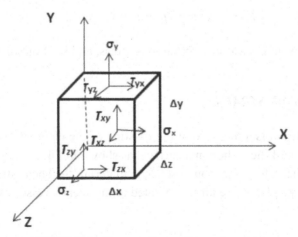

FIGURE 5.4 Stress on three orthogonal surfaces.

This state of stress can be written in matrix form, where the stress matrix [σ] is given by:

$$\sigma = \begin{bmatrix} \sigma x & \tau xy & \tau zx \\ \tau xy & \sigma y & \tau yz \\ \tau zx & \tau yz & \sigma z \end{bmatrix} \tag{1}$$

where, σ = plane stress.

When the stress in one direction, then:

$$\sigma = \begin{bmatrix} \sigma x & \tau xy \\ \tau xy & \sigma y \end{bmatrix} \tag{2}$$

In the case when normal stress σx applied to the element, the direct relationship between stress and strain in the three directions is given by:

$$\varepsilon_x = (\sigma_x \pm v\,(\sigma_y + \sigma_z))/E \tag{3}$$

$$\varepsilon_y = (\sigma_y \pm v\,(\sigma_z + \sigma_x))/E \tag{4}$$

$$\varepsilon_z = (\sigma_z \pm v\,(\sigma_x + \sigma_y))/E \tag{5}$$

If the strain is known, the stress can be given by:

$$\sigma_x = E\,((1-v)\,\varepsilon_x \pm v(\varepsilon_y + \varepsilon_z))/(1+v)\,(1-2v) \tag{6}$$

$$\sigma_y = E\,((1-v)\,\varepsilon_y \pm v(\varepsilon_z + \varepsilon_x))/(1+v)\,(1-2v) \tag{7}$$

$$\sigma_z = E\,((1-v)\,\varepsilon_z \pm v(\varepsilon_x + \varepsilon_y))/(1+v)\,(1-2v) \tag{8}$$

where: E – Young's Modulus N/m^2, σ – Stress, and v – Poisson's ratio.

5.2.2 SHEAR STRAIN

The shear strain is a measure of the twisting of the element, which is the angle change of the orthogonal surface, as shown in Figure 5.5.

For a linear, homogeneous, isotropic material, the shear strains in the γ_{xy}, γ_{yz}, and γ_{zx} planes are directly related to the shear stresses by:

$$\gamma_{xy} = \tau_{xy}/G \tag{9}$$

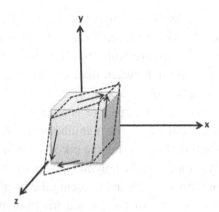

FIGURE 5.5   Shear deformation.

$$\gamma_{yz} = \tau_{yz}/G \tag{10}$$

$$\gamma_{zx} = \tau_{zx}/G \tag{11}$$

For a linear, homogeneous, isotropic material the shear modulus is related to Poisson's ratio:

$$G = E/2(1-2\nu) \tag{12}$$

where: G – shear modulus.

From the above equations, the values of the strength of the material, the specific strength, the Young's modules, Poisson's ratio are the most parameters to compare for the different materials.

5.3 SIMPLIFIED MECHANISM OF FAILURE OF TEXTILE/ POLYMER MATRIX

In the case of textile/polymer we have three zones of materials, as shown in Figure 5.6a, with different mechanical properties. The mechanical characteristics of a fiber/polymer composite depend primarily on the mechanical properties of the combined material of matrix and reinforcement, the surface properties of the fiber, and the nature of the fiber/resin bonding as well as the mode of stress transfer at the interface. Among the many factors that govern the characteristics of composites involving a fibrous material, it is certain that the adhesion between fiber and matrix plays a predominant part, as illustrated in Figure 5.6b, the stress transfer at the interface requires an efficient coupling between fiber and matrix. It is important to optimize the interfacial bonding since a direct linkage between fiber and matrix gives rise to a rigid, low impact resistance material. The treatment of the textile reinforcement during the manufacturing process, will lead to the penetration of the polymer through the porous yarn and bind the fibers together more firmly. Moreover, the polymer may penetrate inside the fiber itself, increasing the yarn strength and reducing the breaking elongation.

The mechanism of failure of the natural fiber/polymer composite is rather complicated comparing to the usual composite, depending on the properties of molded fibers and the matrix. When the permeated fibers become a part of the composite, the final mechanism of composite failure under tension might depend on the following models:

- Applied forces strained the matrix more than the fibers, causing the material to shear at the interface between matrix and fibers.
- Applied forces near the end of the fibers exceed the tolerances of the matrix, separating the fibers from the matrix.
- Applied forces can also exceed the tolerances of the fibers causing the fibers themselves to fracture leading to matrix failure.

Consequently, the reinforcement/matrix interface in composite materials forms in manufacturing processes and determines the performances of the composite. Reinforcements may not be compatible with matrices in view of their physical and/or a chemical property, which causes premature failure of the composites [10]. Usually, coupling materials are used to increase interfacial force between the reinforcement and the matrix. The performance of the adhesion plays a key role in the determination of the bonding strength [11]. The fiber surface treatments are essential to overcome the hydrophilic nature of the natural fibers which can lead to the poor adhesion between fibers and hydrophobic matrix polymer. The natural fibers are inherently incompatible with nonpolar-hydrophobic thermoplastics [12, 13]. The fiber/matrix interface acts an important role in the micromechanical behavior of the composites. The microstructural parameters that control the properties of the composite are: the properties of reinforcement fiber, the properties of matrix, and properties of the fiber matrix at the interface, the fiber volume fraction ratio and their orientations.

5.3.1 COMPOSITE UNDER LONGITUDINAL FORCE

The application of longitude load F causes stress σ_c and strain ε_c in the composite. It is assumed that there is a good bonding of the reinforcement and matrix at the interfacial surface, Zone I, but if not, the mechanism of failure will be completely different and a separation between the fiber reinforcement and the matrix occurs, causing the delamination. The interfacial bonding force usually tested through the pulling of the fiber reinforcement from the composite. The analysis of curve of pulling force vs. displacement will indicate properties of material at Zone II, Figure 5.7. In the proper cases it should be equal to or more than the matrix tensile properties. The shear strength at the interfacial surface determined the suitability of the fiber matrix interfacial bonding. The mechanism of adhesion between fibers and the matrix may be one of the following mechanisms: absorption, chemisorption, diffusion, electrostatic or mechanical interlock or all together depending on the nature of the fiber and polymer.

The crack in the region will start when the stress on the reinforcement is higher than that on the matrix and shearing of the material in Zone II will initiate deboning. That will lead to the beginning of delamination of the matrix through the interface: crack propagation will continue as the stress increases on the composite, reaching a complete separation of the fiber reinforcement from the matrix and the composite will lose its integrity, the load will transfer to the fibers only, Zone III. Figure 5.8 illustrates the difference between good and improper bonding at the interfacial surface.

Natural fiber/polymer failure under longitudinal force F, when there is a complete bonding between the fiber reinforcement and the matrix, is illustrated in Figure 5.9.

Knowing modulus of elasticity of matrix components, then the force at any strain:

$$F = \varepsilon_c \left(E_1 V_1 + E_2 (1-V_1) \right) \tag{13}$$

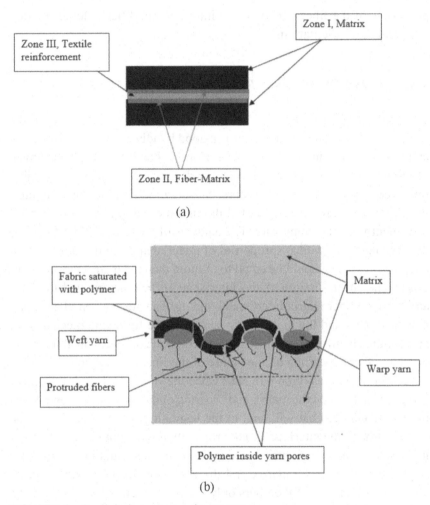

FIGURE 5.6   Textile/polymer composite structure.

where: E_1 – Young's Modulus of the reinforcement, E_2 – Young's Modulus of the matrix, E_3 – Young's Modulus of the interfacial layer between matrix and reinforcement (in this analysis, it is considered $E_3 = E_2$, V_1 – fiber volume fraction, ε_m – Strain in the matrix, ε_f – Strain in the matrix, ε_c Strain in the matrix, σ_c – Stress in the composite, σ_1 – Stress in the fibers,

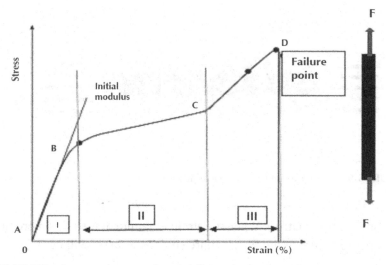

FIGURE 5.7   Stress strain curve of natural fiber polymer composite.

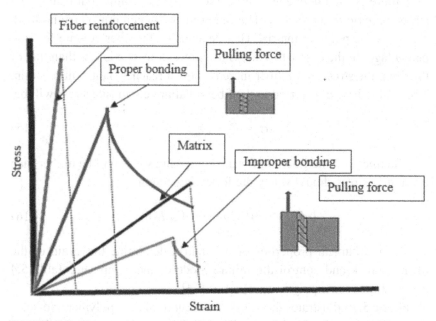

FIGURE 5.8   Stress strain curve of reinforcement and matrix of proper and improper bonding.

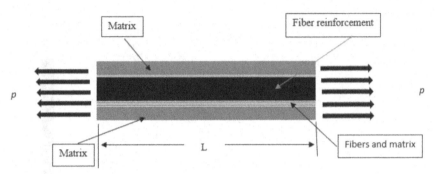

FIGURE 5.9 Composite under longitudinal force.

σ_2 – Stress in the matrix). When the strain to failure of matrix is more than the stress to failure of the fiber, failure composite stress will be:

$$\sigma_c = E_2 \, (1 - V_f) \, \varepsilon_m \qquad (14)$$

Failure of the fibers does not lead to composite failure but results in a stress increase in the matrix. The failed fibers, which now carry no load, can weak the polymer matrix. That depends on the fiber volume fraction percentage. In the case when stress to failure of matrix is less than that of the fiber, the stress in the fiber increases till it reaches their failure strain. The load will be transferred to the fibers. Failure composite stress will be:

$$\sigma_c = E_f . V_f . \, \varepsilon_m \qquad (15)$$

The fiber volume fraction, for the composite stress will be higher than the matrix stress should verify the following conditions;

$$(\sigma_c / \varepsilon_f) \geq ((V_f E_f) + (1 - V_f) \, E_2)) \qquad (16)$$

The mechanical properties of the matrix determine the strain of the fiber at break and control the failure mode of the composite. Table 5.4 demonstrates some polymer properties [14].

Figure 5.10 illustrates the stress strain curve of flax, polymer and composite showing that the failure of flax fibers occurs before the failure of the polymer matrix.

TABLE 5.4 Mechanical Properties of Polymers at Room Temperature [14]

Polymer	Density g/cm³	Young's Modulus E GPa	Yield stress MPa	Breaking stain %	Shear modulus GPa
Polyamide	1.1	0.78	40	5	0.78
Polyester	1.3	3.5	25	2	1.4
Polyethylene	0.95	0.7	25	90	0.42
Polypropylene	0.89	0.9	35	90	0.42
Polyvinyl chloride (PVC)	1.4	1.5	33	55	0.42

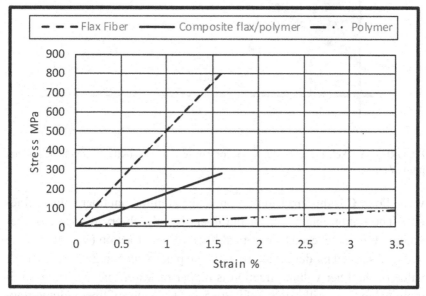

FIGURE 5.10 Stress strain curve for composite and its components.

In the other words, the interaction between the fiber volume fraction, fiber properties and matrix properties plays the distinct role in determining the composite mechanical properties. Figure 5.11 illustrates the relation between the composite stress and the fiber volume fraction.

Stress at points **A** and **D**, are the breaking stress of polymer and fiber material, respectively, **ACD** is the stress in the composite for different fiber volume fraction, Point **B** is the stress in matrix at the fiber breaking strain,

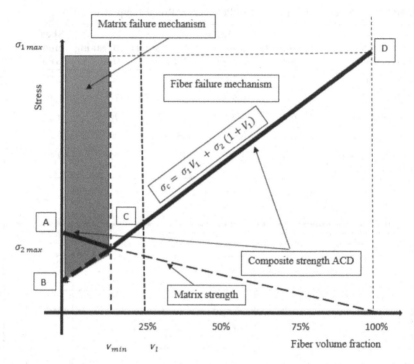

FIGURE 5.11 Schematic illustration of composite stress versus the fiber volume fraction ratio (V1).

while Point **C** is the stress in the composite at minimum fiber volume fraction. The reinforcing action of the fibers is only observed once the fiber volume fraction exceeds the critical fiber volume fraction ($V_1 > V_{min}$). The above described model of failure will take place depending on whether the value of the fiber volume fraction is higher or lower than a certain minimum value V_{min}, as illustrate in Figure 5.11. The value of fiber volume fraction V_l represents the value at which stress in the composite exceeds that the failure stress of the matrix. The value of V_{min} usually is higher than 25% for natural fibers and 2.5% for carbon-polyester composites [15].

5.3.2 COMPOSITE UNDER TRANSVERS FORCE

If a force is applied perpendicular to the composite, then the fibers and matrix will stretch in the same direction, as shown in Figure 5.12.

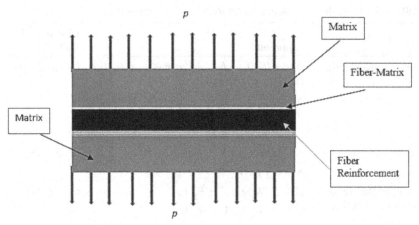

FIGURE 5.12 Composite under perpendicular loading.

The total deformation (Δc) is just the sum of the deformations in the reinforcement ($\Delta 1$), fiber matrix $\Delta 2$ and the matrix ($\Delta 3$):

$$\Delta c = \Delta 1 + \Delta 2 + \Delta 3$$

$$\varepsilon_c = \varepsilon_1 \, V_1 + \varepsilon_2 V_2 \tag{17}$$

And the Young's modulus of the composite will be:

$$E'_c = (V_1/E'_1) + (V_2/E'_2) \tag{18}$$

Then:

$$E'_c = (E'_1 \cdot E'_2)/(E'_2 \cdot V_1 + E'_1 (1 - V_1)) \tag{19}$$

Due to the complicity of the actual cases of composites, several researchers drive a theoretical formula to calculate the mechanical properties of natural fiber polymer matrix [16–22]. Table 5.5 demonstrates a summary of the equations for their approaches.

5.3.3 POISSON'S RATIO OF THE LAMINATE COMPOSITE

The Poisson's ratio of the matrix has the major Poisson ratio "v_{12}" relating to the lateral strain, when a stress is applied in the longitudinal direction

TABLE 5.5 Formulas for Calculations of the Stress in Composite

	Equation	Direction of load
Hirsch's model [16]	$E_c = x (E_1. V_1 + E_2(1 - V_1)) + ((1 - x) E_1. E_2/ (E_2. V_1 + E_1 (1 - V_1)))$	Longitudinal
	$\sigma_c = x (\sigma_1. V_1 + \sigma_2(1 - V_1)) + ((1 - x) \sigma_1. \sigma_2/ (\sigma_2. V_1 + \sigma_1 (1 - V_1)))$	
	x variable $0 \leq x \leq 1$	
Einstein and Guth [16]	$E_c = E_2(1 + 2.5 V_1 + 14.1 V_1)$	Longitudinal
	$\sigma_c = \sigma_2(1 - V_1^{0.667})$	
Modified Bowyer and Bader's model [22]	$E_c = E_1. K_1. K_2. V_1 + E_2. V_2$	Longitudinal
	$\sigma_c = \sigma_1. K_1. K_2. V_1 + \sigma_2. V_2$	
	$0 \leq K_1 \leq 1$ K_2 – Orientation factor	
Kelly and Tyson's mode [16, 20]	$\sigma_c = \sigma_1. \xi_1. \xi_2. V_1 + \sigma_2. V_2$	Longitudinal
	ξ_1 – length efficiency, ξ_2 – orientation factor	
Modified Guth equation	$E_c = E_2(1 + 0.675 S. V_1 + 11.62 S^2. V_1)$	Longitudinal
	S – aspect ratio of the fibers	
Shear-lag equation [19]	$E_c = E_1 (1 - (tanh (0.333 \xi L)/ 0.5 \xi L)) V_1 + E_2. V_2$	Longitudinal
	$\xi = (2 E_2/(E_1 (1 + vm). ln(P_f/V_1),$	
	P_f – fiber packing factor	
Cox-Krencel [18]	$\sigma_c = E_1. \xi_1. \xi_2. V_1 + E_2. V_2$	Longitudinal
	$\xi_1 = (1 - (tanh(0.5 \beta. l_f)/(0.5 \beta. l_f))$	
	$\beta = (2/d_f) (2 G_m/(E_1. ln ((\pi/X_1. V_1)^{0.5})$	
	$G_m = E_2/2(1 + v_f)$	
Modified role of mixture [20]	$\sigma_c = \sigma_1. \beta. V_1 + \sigma_2. V_2$	Longitudinal
	β – additional coefficient which accounts for the weakening of the composite due to fiber orientation and is less than 1	
Brintrup [17, 20]	$E_c = E_1. E_2/(E_1(1 - V_1) + E'_2 V_2))$	Transverse
	$E'_2 = E_2/(1 - v^2)\pi$	
Halphin–Tsai [17, 20]	$E_c = E_2((1 + 2\beta V_1)/(1 - \beta V_1))$	Transverse
	$\beta = ((E_1/E_2) - 1)/((E_1/E_2) + 2)$	
	$\sigma_c = \sigma_m ((1 + 2 \beta_1. V_1)/(1 - \beta_1 V_1))$	
	$\beta_1 = ((\sigma_1/\sigma_2) - 1)/((\sigma_1/\sigma_2) + 2)$	

TABLE 5.5 (Continued)

Equation	Direction of load
Chen's Equation [21] $\sigma_0 \cos^2\theta = \beta \sigma_L, 0 \leq \theta \leq \theta_1, \theta_1 = tan^{-1} (\tau_m/\sigma_L)$ τ_m is the shear strength of the matrix, σ_0 is the strength of composite of all aligned at θ to the direction of applied stress, σ_L is the strength of the composite with all the fiber aligned in the direction of applied stress, β is the strength efficiency factor which relates the strength of a discontinuous fiber composite to the strength of a corresponding continuous fiber composite.	Random fiber composites

and "v_{21}" Poisson's ratio relating to the strain in the longitudinal direction when a stress is applied in the traverse direction, $v_{12} > v_{21}$. Strain ε_2, when a stress is applied in the longitudinal direction, divided by the longitudinal strain ε_1, $v = -\varepsilon_2/\varepsilon_1$. Consequently,

$$v_{12} = v_f.V_1 + v_m. (1 - V_1) \tag{20}$$

and

$$v_{21} = v_{12}(E'_2/E'_1) \tag{21}$$

5.3.4 THE MINIMUM VALUE OF FIBER VOLUME FRACTION

In all the above equations, the fiber volume fraction performs a significant role in determining the overall matrix stiffness under parallel or perpendicular loading. The effect of fiber volume fraction is more pronounced in the case of parallel loading. Analysis of the modes of failure, indicates that there are several other situations of failure due to the complex structure of the natural fiber composites, which can be summarized as:

1. The inhomogeneity of the natural fiber physical and mechanical properties;

2. Irregularity of the yarns in diameter and number of fibers in the cross section;

3. Heterogeneity of the properties of the fiber mat forming the reinforcement;

4. Volume fraction of the fiber variation;

5. The orientation of the fibers.

The above analyses specify the importance of studying the volume fraction of the reinforcement.

The natural fiber/polymer composites are heterogeneous material hence it consists of completely different materials in mechanical, thermal, and electrical properties. The reinforcement has high stiffness and strength, the matrix has low stiffness and strength. Thus, when both are subjected to the same load, it results in the unequal distribution of the stress in the same cross section of the composite, and there are two modes for failure. One mode is the matrix failure mode and the other fiber failure mode. Both modes depend on the fiber volume fraction and the properties of the matrix components.

For longitudinal force in the direction of the reinforcement, the force on the composite will be equal to:

$$\sigma_c Ac = \sigma_1 A_1 + \sigma_2 A_2 \tag{22}$$

where: σ_c = composite stress; σ_1 = reinforcement stress; σ_2 = matrix stress; A_c = composite cross sectional area; A_1 = reinforcement cross sectional area; A_2 = matrix cross sectional area; E_1 = Elastic modulus of the fibers in longitudinal direction; E'_1 = Elastic modulus of the fibers in transfer's direction; E_2 = Elastic modulus of the matrix in longitudinal direction; E'_2 = Elastic modulus of the matrix in transfer's direction;

The equation rearranged to give the rule of mixture for longitudinal stress:

$$\sigma_c = \sigma_1 V_1 + \sigma_2 V_2 \tag{23}$$

Assuming the component of the matrix is isotropic, and the strain in all parts of the composite is the same, then:

$$E_c = E1 V_1 + E_2 V_2 \tag{24}$$

$$\sigma_c = \sigma_1 V_1 + \sigma_2 (1-V_1) \tag{25}$$

In design of the composite the value of resultant σ_c should be greater than σ_2:

$$V_1 = (\sigma_c - \sigma_2)/(\sigma_1 - \sigma_2) \tag{26}$$

Assume $R = (\sigma_c/\sigma_1)$ and $R_1 = (\sigma_2/\sigma_1)$, then the fiber volume fraction for getting a certain composite force F_c:

$$F_c = A_c \cdot \sigma_c \tag{27}$$

$$F_c = A_c (R - R_1)/(1 - R_1) \tag{28}$$

For example, if the composite failure stress is 0.7 that of fiber failure stress, then $R = 0.7$, and if $R_1 = 0.25$, then we need fiber minimum volume fraction V_1 be 0.6.

In all the above, the fiber length is assumed higher than the critical fiber length, which can be calculated from the following equation:

$$l_c = \sigma_1 d_f/2\tau \tag{29}$$

where l_c – critical fiber length (for continuous fibers $L \gg 15\ lc$, while for short fibers $L \leq 15\ l_c$); τ – interfacial shear stress (Table 5.4 gives the shear modulus of some polymers).

The value of the modulus of elasticity can be calculated as:

$$E_c = E_1 V_1 + E_2 (1 - V_1) \tag{30}$$

For $E_1/E_2 = 100$, volume fraction V_1 be 0.6, then $E_c/E_1 = 0.604$.

The ratio of reinforced bearing force to the composite force can be expressed as:

$$(P_1/P_c) = V_1/(V_1 + (E_2/E_1) (1 - V_1)) \tag{31}$$

The value of the fiber volume fraction V_1 depends on the requirement of the composite design to bear a certain value of load P_c:

$$(P_c/P_l) = (V_l + (E_2/E_l) (1 - V_l))/V_l \qquad (32)$$

Usually $E_l \gg E_2$, hence the ratio $(E_2/E_l) = R_2 > 0.01$ to 1, and the ratio $(P_c/P_l) = R_3$ is depending on the fiber volume fraction.

Figure 5.13 specify that the fiber loading ratio in the composite depends on the elasticity modules ratio of fibers to matrix and the fiber volume fraction ratio. In all cases, the effect of fiber volume fraction ratio parameters is reduced after $V_l > 0.3$ and the elasticity modules ratio of fibers to matrix becomes the dominant factor. For $R_2 > 0.02$ higher fiber volume fraction is required.

It is worth to mention that the actual volume fiber fraction will be less than the calculated due to the matrix porosity. Moreover, with high value of fiber volume fraction the spaces between the fibers become smaller and it is expected to have a bad transfer of the stress between the fiber and matrix; the shear stress increasing leads to the delamination between the fibers. It is recommended for long aligned fibers higher values of volume fiber fraction be up to 60% and for short randomly fibers, lower values are up to 30%.

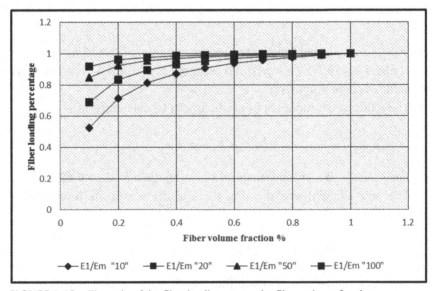

FIGURE 5.13   The ratio of the fiber loading versus the fiber volume fraction.

5.4   CALCULATIONS OF FIBER VOLUME FRACTION OF NATURAL/POLYMER COMPOSITE

NFPC composite may consist of loose fiber aggregate compressed in the thin layer and for short fibers, Figure 5.14, it will be randomly laid. However, for long fibers, they may be laid unidirectional or multi-directionally in a several layers. Yarns may also be used as spun yarns parallel in one direction or orthogonally laid or they may be woven or knitted or braided to form a fabric in 2-D or 3-D. using the fibers as reinforcement, they occupy the largest fraction of the composite and bear the major portion of the load. Consequently, the fiber volume fraction and fiber orientation play important role of the final composite characteristics. Usually, the synthetic fiber is different from the natural fiber in many characteristics, such as density, moisture absorption, cross section shape, and fiber morphology. In the case of NFPC composite, the matrix combines the fibers together, holding them aligned in the important stressed directions. Loads applied to the composite are then transferred into the fibers, the principal load-bearing component, through the matrix, enabling the composite to withstand compression, flexural and shear forces as well as tensile loads. The ability of composites reinforced with short fibbers to support loads of any kind is dependent on the presence of the matrix as the load-transfer medium, and the efficiency of this load transfer is directly related to the quality of the fiber/matrix bond. The matrix must also isolate the fibers from each other so that they can act as separate entities. In all these structures the volume of fiber architect will depend on the pressure applied on it during the manufacturing of the composite.

FIGURE 5.14   Random fiber architect in composite.

5.4.1 FIBROUS STRUCTURES

Several investigators indicate that the fiber volume fraction is one of the main parameters in the determination of the NF composite mechanical properties, consequently we need to determine its exact value. Generally, fiber volume fraction is calculated as:

$$V_f = \rho_m W_f / (\rho_m W_f + \rho_f W_m) \qquad (33)$$

where: V_f – volume fraction of fibers; W_f – weight of fibers, W_m – weight of matrix, ρ_f – density of fibers, ρ_m – density of matrix.

The natural fibers are affected by the absorption of moisture in two directions: it changes the fiber density through the weight of water absorbed as well as the swelling of fiber itself. The real fiber weight W_f should be modified by W_{f1}:

$$W_{f1} = W_f (1 - W_c) \qquad (34)$$

where: W_c – weight of water content, ρ_f – density of fibers, and ρ_{f0} – density of dry fibers.

The value of measured fiber density ρ_f should be corrected, according to the value of the moisture content M_c:

$$\rho_{fc} = (1 + M_c) / ((1/\rho_{f0}) + M_c) \qquad (35)$$

Hence, the Eq. (33) will be:

$$V_f = \rho_m W_{fc} / (\rho_m W_{fc} + \rho_{fc} W_m) \qquad (36)$$

Figure 5.15 shows how the fiber density will change with the moisture content depending on the dry fiber density. Thus, the moisture content problem should be taken into consideration when estimating the fiber volume density. When manufacturing natural/polymer composites, it is recommended to use as dry fibers as possible.

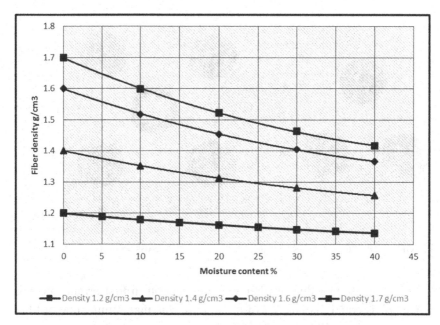

FIGURE 5.15 Effect of the moisture content on the fiber measured density.

5.4.2 IDEAL FIBERS ASSEMBLY

General situation of packing of the fibers depend on the structure in which the fibers are put in. It will be different if the fibers are parallel to longitudinal direction of the matrix or random, or in the yarn which forming a fabric in a predetermine space location. Assuming the fiber structure consists of fibers of circular cross section of diameter "d," two different packing arrangements were considered, square and hexagonal close packing [23–25]. Hexagonal close packing has a higher packing density. The fiber arrangement of this type is schematically shown in Figure 5.16. Considering the hexagonal element and according to the definition of fiber volume fraction of a composite, we have the maximum fiber volume fraction in this case is equal to 0.907. The packing density of maximum fiber volume fraction in this case cubic packing is equal to 0.786 [26].

Fiber volume fraction:

$$V_f = (0.9075 \ (d/p)^2) \tag{37}$$

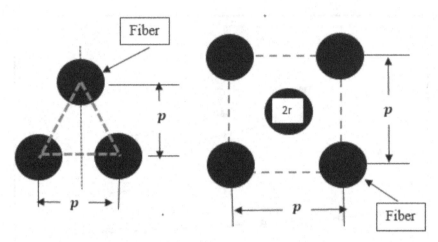

FIGURE 5.16   Idealized fiber arrangement in composite cross section.

The maximum value of fiber volume fraction in first arrangement is when "p" is equal to diameter of the fiber d. The maximum value of fiber volume fraction will be equal to 0.907.

While in the square arrangement, Figure 5.16, the value of the fiber volume fraction will be:

$$V_f = (0.785 \ (d/p)^2) \tag{38}$$

The maximum value of fiber volume fraction is when $P = d$ and equal to 0.785.

The practical value of V_f is varied between 0.5 and 0.8. There are many factors affecting its practical value, such as:

1. Fiber length
2. Fiber aspect ratio
3. Fiber rigidity
4. Yarn count
5. Yarn twist
6. Yarn unevenness
7. Weave structure
8. Fibrous structure of mat

The yarns take the multi-directional form through weaving, knitting, braiding, and 3-D fabrics. Because the reinforcement is oriented in 3-D space, the value of maximum fiber volume fraction is reduced below the theoretical value.

When a matrix is formed of random orientated fibers, the value of volume fraction decreases depending on the spacing ratio R/r. In this case, the fiber volume fraction may be given by:

$$V_f = (r/R)^2 \, V_{fmax} \qquad (39)$$

where: fiber spacing R is the radius of the composite, r is the fiber radius.

In the case of random fibrous structures (R/r) may reach 0.10–0.30. Figure 5.17 reveal the fiber volume fraction versus the ratio (R/r), which indicates that as the fiber becomes finer, the fiber volume fraction reduces. Figure 5.17 shows that, the V_f reaches its maximum value when R is equal to r and will decrease rapidly with the increase of (R/r). Typical fiber volume fractions are only 20% to 40% for NFPC [40]. As shown in Figure 5.18, the value of (R/r) in some areas may be higher than 3 to 5.

The shape of fiber cross sectional area also affects the fiber volume fraction, the round cross section has the less volume fraction. For instance,

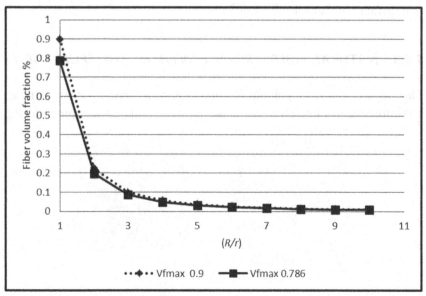

FIGURE 5.17   Fiber volume fraction versus the spacing ratio (R/r).

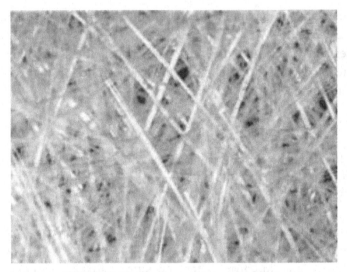

FIGURE 5.18   Random arrangement of fibers

if we have two cross sections, as shown in Figure 5.19, the square one has higher fiber volume fraction than the round fiber cross section.

The practical values of the fiber volume fraction can reach for flat 'non-crimp' yarns 0.6, for woven fabric 0.4–0.55 and for random fiber arrangement in mat form 0.15–0.25.

5.4.3   NATURAL FIBER ARRANGEMENT IN THE SPUN YARNS

The value of the fiber volume fraction is not the same for spun yarns as a reinforcement material to form the NFPC composite compared to the

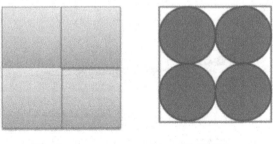

$Vf = 1$ $Vf = 0.789$

FIGURE 5.19   Effect of the shape of fiber cross section.

continues yarn arrangement due to the many reasons; the fibers are varied in diameter and cross section shapes as well as rigidity, the fibers are not straight and parallel but they migrate between the cross section of the yarn along its length, yarn is twisted to get its integrity which will exert a radial pressure on the different fibers in the yarn cross section, make the fiber's packing density varied and, consequently yarn packing density, also each spinning system produces yarns of different structures. Figure 5.20 expose a cross section in a spun yarn.

The above analysis of the real yarn packing density indicates that it is essential to determine the value of fiber volume fraction for the real yarn. Fiber packing density of idealized yarn structures and the distribution of the fiber in the yarn cross section were studied by several investigators [26–36]. According to various authors, the packing density may be defined as: The ratio of the volume of fibers to the volume of yarns, the ratio of yarn density to fiber density, the ratio of total cross-section area of fibers to the cross-sectional area of the yarn. Assuming the yarn is divided into rings of radius r_i so that the areas of such rings are equal, then average value of fiber volume fraction of yarn is:

$$V_f = (\Sigma \ (A_{fi}/A_i)/m) \tag{40}$$

For $i = 1, ..., $ m.

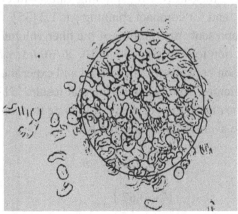

FIGURE 5.20   Cross section of spun yarn in composite.

where: m – number of the rings, A_{fi} – area of fibers in the ring of ith which has an area A_i.

When the values of the radial packing density for a certain yarn are equal to 0.7, 0.6, 0.5, 0.42, and 0.25, respectively, for five consecutive rings, the average fiber volume fraction will be 0.494. In this analysis we assume that the yarn diameter is circular and constant along its axis. For theoretical determination of yarn fiber volume fraction, a number of different model forms were tried. Fiber distributions have been studied by a several researchers, the parabolic model [26–31] fits a wide variety of yarn types, including hollow-centered bundles used in the production of yarns. For yarns that are not hollow-centered, a simplified form of the parabolic equation is:

$$V_f(r) = V_{fmax} \left(1 - (r/R)^2\right) \tag{41}$$

In the idealized structure, the maximum value of V_{fmax} has been proved to vary between 0.785 and 0.887 [27], depending on the arrangement of the fibers. In the case of real yarns, we can consider V_{fmax} at the center of the yarn and there is minimum number of fibers n_{min} required for its formation [36, 38].

$$V'_{fmax} = \gamma. \, (n_{min}/n_y)^{0.5}. \, V_{max} \tag{42}$$

where: $n_y = tex_y \times 10/dtex_f$, γ is a factor depending on the spinning system and assumed to be: for carded ring spinning value will be 1, for combed ring spinning 1.1, and for compact spinning γ is 1.2 [35].

Figure 5.21 represents distribution of the fiber volume fraction across the yarn cross section for yarns of counts 10, 20, 30 *tex*, while Figure 5.22 shows a comparison between the theoretical and experimental results that are in a good matching with the experimental results [31]. From the Eq. (40) we can get the distribution of the fiber volume fraction across the yarn cross section. The average value is given by:

$$V_f = \int V_f(r) \, dr \, / \int dr \tag{43}$$

$$V_f = 0.67 \, V_{fmax} \tag{44}$$

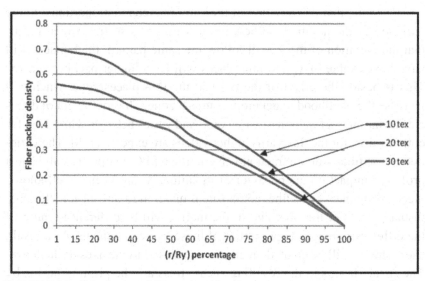

FIGURE 5.21 Distribution of fiber volume fraction across the yarn cross section for ring spun yarn counts 10, 20 and 30 *tex*.

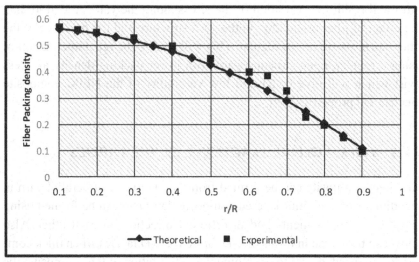

FIGURE 5.22 Theoretical and experimental distribution of radial fiber volume fraction for ring spun yarn count 10 *tex*.

Assuming that the variation of the yarn diameter is proportional to the variation of the number of fibers in the yarn cross section which means that the variation of the V_f will follow the same pattern, thick places will have lower value of V_f and thin places will have higher packing density. This is beside the effect of the twist in the thin places zone which will increase the likelihood to get more compact zone. The coefficient of diameter variation is a function of yarn count and varies between 8 and 14% for cotton yarns, for the flax yarns higher values are expected [38]. The variation of the fiber volume fraction value in the NFPC composite will have a profound impact on the behavior of its failure. It has been a well-known mechanism that when a fiber composite is under the axial tension, the axial displacements in the fiber and in the matrix will be different because of the differences in tensile properties of these two components. As a result, shear strains will be created on all planes parallel to the axes of the fibers. The shear strain and the resulting shear stress are the primary means by which load is transferred to fibers (for a short fiber composite), or distributed between and supported by the two components of composites. Interaction between fibers and matrix means that fiber reinforcing function is realized [26]. Therefore, it is expected that variation of the fiber volume ratio will result in a variation of the stress along the yarn. Added to that, the effect of fiber orientation in the yarn structure and the presence of the fiber hairiness are not as regular. This will lead to the recommendation of using a regular, open structure yarn with lower packing density and less twist which will give better and low variability of the NFPC composite mechanical properties [39].

5.4.4   FIBER VOLUME FRACTIONS IN THE FABRICS

Architect composite can be formed from one laminate of parallel yarn in one dimension and multilayered composite laminate can be formed using one dimensional laments laid in different direction on each other. Also yarns can be woven in forms of 2-D fabrics or 3-D fabrics when thick composites are needed, and to overcome the delimitation of the composite, the weave of the yarns may be orthogonal (pain weave) multidirectional fabric, weft knitted, warp knitting, biaxial braided fabric and tri-axial braided fabric. The yarns with a fiber volume fraction and when they are architected in a certain structure, the total fiber volume fraction will depend upon the spaces between the yarns in the fabric form. The determination of the area between yarns in the fabric is not sufficient to calculate the fiber volume fraction since the yarns will be filled by the matrix material. Consequently, the fiber volume fraction will be:

V_f = (Volume of the fiber in the yarns/Whole assembly volume)

$$V_f = (\rho_m \cdot W_y / (\rho_m \cdot W_y + \rho_y \cdot W_m)) \tag{45}$$

where: ρ_y – yarn density, ρ_m – matrix material density, W_y – weight of yarns, W_m – weight of matrix. Volume fiber fraction of weave structure is the percentage of the volume of the fiber to the volume of the composite.

In practice, the different components of the composite percentage are usually measured by weight, W_i consequently their sum will be:

$$\Sigma_1^n W_i = 1 \tag{46}$$

where, n is the number of the component partial weight fraction.

Similarly, the partial volume fraction of the composite of the components V_i will be:

$$\Sigma_1^n V_i = 1 \tag{47}$$

where, n is the number of the component partial volume fraction.

Transfer from Eqs. (46)–(47), knowing the density of each component of the composite, the following equations can be applied:

$$V_{fl} = (W_{fl}/\rho_{fl})/(\Sigma_1^n (W_i/\rho_i)) \tag{48}$$

$$W_{fl} = (V_{fl}/\rho_{fl})/(\Sigma_1^n (V_i \cdot \rho_i) \tag{49}$$

The composite density can be given as:

$$\rho_c = \Sigma_1^n (V_i \cdot \rho_i)/V_c \tag{50}$$

where: V_c – volume of composite.

5.4.5   VOIDS VOLUME FRACTION

During the processing of the composite, a several voids between the reinforcement structures or in the matrix layers may occur. This reduces the
matrix volume fraction directly and also affects the bonding of the reinforcement in the matrix, thus leading to the delamination due to stress
concentration. The volume of voids ratio is given by:

$$V_R = V_v/V_c \tag{51}$$

where: V_v – volume of voids, and V_c – volume of composite.

The number of voids and their distribution in composite affects its
strength. If V_v is the void content, then the Eq. (45) for simple composite
becomes:

$$\rho_c = \rho_f V_f + \rho_m (1 - V_f - V_v) \tag{52}$$

The exact theoretical value of the fiber volume fraction for the fabrics
has been studied by several investigators [41–44], especially when continuous synthetic yarns are used for the formation of the fabric. However,
it is not regarding the natural fiber yarns with their variability. Thus, it is
recommended to measure the porosity of the fibrous structure in order to
determine matrix volume. Porosity and air permeability have a significant
correlation. The measured value of the air permeability of the fabric or
any fibrous structure can be used to calculate the structure porosity, both
vertical and horizontal [43]. The practical values of the fiber volume fraction depend on the structure of the reinforcement and its porosity. The
structures with low porosity allow high volume fraction. Practical value of
fiber volume fraction for unidirectional yarns is 50–70%, for woven fabrics 35–55%, and nonwoven mat 10–30%. The void size percentage of the
voids and their distribution have different effects on the stress at failure.

5.5   LAMINATE AND HYBRID COMPOSITE

In order to reach the requirement of the composite design, more than
one laminate might be needed to withstand the stresses applied on the

composite. The laminate may be made of the same laminate's material inlayer one over the other, however there are several techniques to design the composites:

1. Through the change the axis directions of the laminates (Orientation).
2. Hybrid composites.

The hybrid composite may be made of different laminates of various materials and structures or the materials used for laminates are a hybrid material. Figure 5.23 shows some types of Hybrid composite.

The fiber volume fraction referred to laminate and hybrid composite can be calculated using fiber volume (V_f) of each component, thickness (t) of the composite, fiber weight percentage of each component (W_i), total weight of the composite (W_c), so the total fiber volume fraction is:

$$V_f = (W_c/t) \, (\Sigma_1^n \, (W_i/\rho_i)) \tag{53}$$

where: ρ_i is fiber density for each of n fiber type.

5.5.1 SIMPLE APPROACH TO STRESS IN THE MULTI LAMINATE COMPOSITE

Composites are usually weaker in direction transverse to the fiber direction [45]. In the case of multi laminate composite, the direction of main fiber axis is rotated by an angle Θ relative to the longitudinal direction of the composite.

From Figure 5.24 the equation can be derived [45]. The stress in the fibers due to stresses σ_{cx} and σ_{cy} applied on the composite will be:

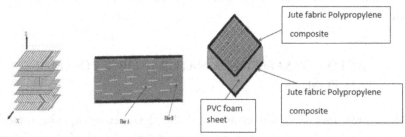

Laminates Composite   Hybrid Composite

FIGURE 5.23 Hybrid composites.

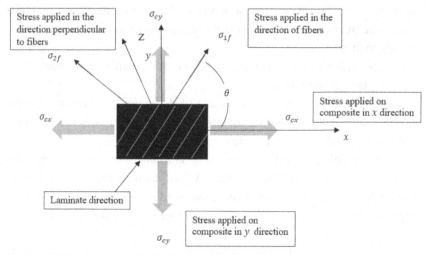

FIGURE 5.24   Stress in multi-laminate composite.

$$\sigma_{1f} = \sigma_{cx} \cos^2\theta + \sigma_{cy} \sin^2\theta + 2\,\tau_{xy} (\cos\theta + \sin\theta) \qquad (54)$$

$$\sigma_{2f} = \sigma_{cx} \sin^2\theta + \sigma_{cy} \cos^2\theta - 2\,\tau_{xy} (\cos\theta + \sin\theta) \qquad (55)$$

$$\tau_{12} = -\sigma_{cx} \sin\theta.\cos\theta + \sigma_{cy} \cos\theta.\sin\theta + \tau_{xy} (\cos^2\theta - \sin^2\theta) \qquad (56)$$

where: σ_{1f} – stress to the fiber axis, σ_{2f} – stress perpendicular to the fiber axis.

The analysis of composites with fibers inclined or oriented with respect to the axial contribute low value to the stress of the composite, depending on the angle of alignment θ. The total laminate composite stress will be the sum of the share of each laminate in the direction of the applied force.

5.6   FACTOR OF SAFETY FOR NATURAL FIBER POLYMER COMPOSITES

The safety factor should ensure that the designed composite material is safe under the specified extreme conditions and provide confidence against collapse, taken into consideration for unavoidable differences between assumed and actual structural behavior. During the design processes the material properties are measured under different conditions than that when the material is subjected during life cycle. Moreover, the properties of natural fibers may be changed with utilization or due to the effect of environmental conditions.

The structure for analysis will lead to the error in the evaluation of the structure behavior under actual load. Surrounding working environment have a profound effect on material properties. Such factors will lead to the uncertainty in the design data and consequently, require the increase of the material and/or the design capacity to acceptable risk level. Another factor, which increases the value of factor of safety of the design, is if the element failure will cause damage of equipment or loss of life. It should be mentioned that the choice of the value of the factor of safety will affect the design cost. Aspects effecting factor of safety are:

- Type of Loading: static, dynamic direct tension or compression, direct shear, bending, torsional shear
- Nature of load application: uniaxial, biaxial, triaxial, types of stresses, variations of loads over time, repeated & reversed, fluctuating, shock or impact
- Material: properties, ultimate strength, yield strength, endurance strength, ductility
- Confidence: reliability of data for loads, material properties, stress calculations
- Cost: value of the design factor
- Environment: moisture conditions, UV effect, temperature, chemical exposure.

5.6.1 DEFINITIONS

Factors of Safety: Multiplying factors to be applied to limit loads or stresses for purposes of analytical assessment (design factors) or test verification (test factors) of design adequacy in strength or stability [46]

Limit Load: The maximum anticipated load, or combination of loads that a structure may experience during its design service life under all expected conditions of operation

Margin of Safety (MS): MS = [Allowable Load (Yield or Ultimate)/ Limit Load Factor of Safety (Yield or Ultimate)]

Global Safety Factor: The factor of safety can be applied either to the loads (the "load factor" method) or to the material strengths (the "permissible stress" method).

Limit State Design: Some partial safety factors are applied to the load and others to the material strengths. The structural design must meet two "limit states": the Ultimate Limit State and the Serviceability Limit State.

a) The Ultimate Limit State requires that the structure must withstand the highest applied load without collapsing catastrophically.

b) The Serviceability Limit State requires that the structure must not suffer excessive deflection, cracking, fatigue, vibration, fire damage or other degradation under its normal working conditions. The partial safety factors for this limit state are lower.

5.6.2 PROBABILISTIC METHODOLOGY

According to the application of the composite, there are several approaches to determine the value of the factor of safety [46–49], such as the probabilistic approach, failure mechanisms analysis, global safety factor. In all these approaches the factor of safety is a combination of partial factors that connected with the materials and the others connecting with the load applied on the structure and the level of uncertainty in the measured or calculated values at a certain acceptable level of failure risk.

The probabilistic approach can be used for the determination of the factor of safety for a certain design. First the acceptable risk level should be defined according to product of failure probability. The probability of failure of the material under load "P" is equal to the area under the normal distribution curve as shown in Figure 5.25. However, the composite may be subjected to several types of loads, each has its failure mechanisms. Based on evaluating all possible failure mechanisms separately, each failure mode has its own failure mechanisms, such as fiber failure, delamination, yielding. In a complete analysis all failure modes and mechanisms need to be evaluated in detail.

The largest influence to the safety factor comes from the scatter of material's mechanical properties, quantified by coefficient of variation

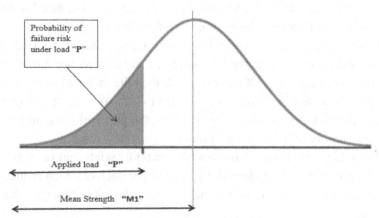

FIGURE 5.25 Failure probability under constant load.

(CV). The second important factor is the variability of the load during use of the composite material.

The variability of the material properties may be assumed to follow normal distribution of mean and M1 and standard deviation σ1. The probability of failure of the martial under load P is equal to the area under the normal distribution curve as shown in Figure 5.26. Assuming the mean

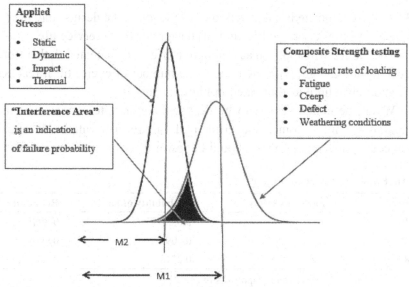

FIGURE 5.26 Failure probability under variable load.

M1 is 100 N and standard deviation 30 N of the composite, then for different applied loads the reliability of use is presented in Table 5.6.

If the applied load has variable values, Figure 5.26, the probability of failure will differ; assume the load is varied and follows normal distribution curve of mean M2 = 80 and σ_2 = 10, then the factor of safety is 1.25 and the probability of failure will increase to 0.2586. Statistical approach defines the factor of safety that insures that the part will work under load with probability of 0.99 without failure.

The other aspects that determine the factor of safety can be expressed in the form of multiplication of the several parameters which reduce the laboratory measured strength value σ_u; the allowable strength can be expressed as:

$$\sigma_{allowable} = factor\ of\ safety.\sigma_u \qquad (57)$$

where: *factor of safety* = $K_1.K_2.K_3.K_4.K_5$, K_1 – manufacturing factor, K_2 – size factor, K_3 – load factor, K_4 – temperature factor, and K_5 – reliability factor.

5.6.3 FACTOR OF SAFETY FOR FATIGUE LOADING

When the composite is subjected to cyclic loading, the design may be: (i) design for infinite service life, and (ii) design for finite service life.

The fatigue will apply an accumulative effect on the material, therefore reducing its strength with increase of the number of cycles till it reaches the endurance limit as illustrated in Figure 5.26.

When the component is to be designed for infinite service life, the endurance limit becomes the criteria of failure. The amplitude stress induced in such components should be lower than the endurance limit in

TABLE 5.6 Probability of Failure

Force	Factor of safety*	Probability of failure	Reliability
10	10	0.0.0013	0.9987
30	3.3	0.0099	0.9901
80	1.25	0.2514	0.7486

*Factor of safety = (failure stress/applied stress).

order to withstand the infinite number of cycles. Applying the following equation:

$$\sigma_{allowable} = (\sigma_{endurance} / \textit{factor of safety}) \tag{58}$$

5.6.4 CREEP–FATIGUE LOADING

Creep is a time dependent deformation following application of a constant load to a textile material and polymer. Fatigue failure occurs when a material ruptures at a much lower stress compared to its breaking stress under cyclic deformations on continued usage. Creep and fatigue of textile reinforcing materials should be taken into consideration for developing a high performance composite [50]. Severe creep type deformation can occur under cyclic loading provided the combination of stress and temperature. Composites are more creep resistant than the matrix due to the introduction of fibers into the matrix. Creep is dominated by the creep behavior of the matrix in composite materials. The load carried by matrix will be outbuilding as the matrix deforms in creep, that load will be shared by the intact fibers which can be assumed not to creep. More and more load will be carried by fibers as creep continues. Finally, the entire load will be carried by fibers when matrix stresses are relaxed during the creep [51]. When a relatively large load is applied, a substantial number of fibers can fracture and initiate damage during the initial loading and creep of the matrix. In addition, some defects introduced during the manufacture of fibers can also cause fracture. As fibers break, the load on the cross-section is redistributed. The load shedding from the broken fibers is shared by the intact fibers and the matrix, which increases the creep strain and causes more fibers to break.

A relationship between the creep life and stress level has been suggested [52] to be used for creep life prediction for the composite materials under constant loading.

$$T_c = A\sigma^{-n} \tag{59}$$

where: T_c is the creep life, σ is the stress level, "A" and "n" are the empirical constants that are material related.

Fatigue damage and creep damage increase with applied loading cycles in a cumulative manner that may lead to fracture. At elevated temperatures, the processes of fatigue damage (due to cyclic loading) and creep damage (due to the loading duration) can also interact [52]. The factor of safety in the case of creep–fatigue loading should be increased. The value of $\sigma_{endurance}$ is definitely decreased, depending on type of matrix polymer and temperature, as illustrated in Figure 5.27. When composite parts are subjected to high temperature thermal cycles concurrently with mechanical strain cycles, conditions resulting in microstructural damage occur. That will increase the creep strain as the lifetime of the components under such loading is found to be quite different from that obtained in fatigue – creep damage of the material at low temperature [53].

5.6.5 PRACTICAL VALUES OF THE FACTOR OF SAFETY

The nature of the working load is the important factor in determining the factor of safety, represented in the determination of maximum steady-state load, peak dynamic load, beside to breaking load of composite structure, elastic limit of composite structure, testing methods of composite

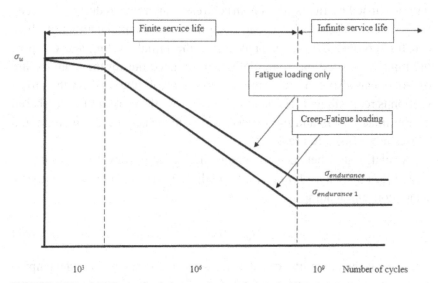

FIGURE 5.27 S-N Curve for fatigue and creep-fatigue loading.

material, effect of environment. Several approaches are given for estimating the factor of safety which may vary between 2 and 6, depending on the above stated conditions of the applied loads and the material used.

KEYWORDS

- **choice of matrix**
- **composite design**
- **creep–fatigue loading**
- **factor of safety**
- **material selection**
- **mechanism of failure**
- **Poisson's ratio**
- **stress analysis**
- **volume fraction**

REFERENCES

1. Mansor, M. R.; Hambali. A.; Azaman M. D. Material selection of thermoplastic matrix for hybrid natural fiber/glass fiber polymer composites using analytic hierarchy process method. Proceedings of the International Symposium on the Analytic Hierarchy Process 23–36 June, 2013, Kuala Lumpur, Malaysia.
2. Saba, N.; Tahir, P.; Jawaid, M. A Review on Potentiality of Nano Filler/Natural Fiber Filled Polymers composites. Polymer. [Online] 2014, 6, 2247–2273 file:///C:/Users/magdy/Downloads/polymers-06-02247-v2%20(2).pdf (accessed May 5, 2016).
3. Bekaert co. http://www.bekaert.cm/ (accessed May 5, 2016).
4. Cherouat, A.; Borouchaki, H. Present State of the Art of Composite Fabric Forming: Geometrical and Mechanical Approaches, Materials. [Online] 2009, 2, 1835–1857 file:///C:/U2016sers/magdy/Downloads/materials-02–01835%20(7).pdf (accessed May 5, 2016).
5. STIG, F. 3D-woven Reinforcement in Composites, PhD. Thesis, KTH School of Engineering Sciences, Stockholm, Sweden 2012.
6. STIG, F. An Introduction to the Mechanics of 3D-Woven Fiber Reinforced Composites, PhD. Thesis, KTH School of Engineering Sciences Stockholm, Sweden 2009.
7. Unal, U. P. G. Classification of 3-D woven fabrics, Woven Fabrics, Chapter 4. [Online] May 16, 2012.Edited by Han-Yong Jeon, ISBN 978-953-51-0607-4, Publisher: InTech http://www.intechopen.com/books/woven-fabrics (accessed May 15, 2015).

8. El Messiry, M.; Manday, A. Investigation of Knitted Fabric Dynamic Bagging for Textile Composite Preforms. JT I. 2016, 107(4), 431–444.

9. Raju, J.; Manjusha, S.; Duragker, N.; Jagannathen, N.; Manjunatha, C. Prediction of onset of mode I delamination growth under a tensile spectrum load. J. Mater. Sci. Res. 2014, 3(2), 45–51.

10. Madell, J. F.; Chen J. H.; Mc Garry F. J. A micro-debonding test for in situ assessment of fiber/matrix bond strength in composite materials. International Journal of Adhesion and Adhesives July, 1980, 1(1), 40–44.

11. Nishikawaa, M.; Okabeb, T.; Hemmia, K.; Takedac, N. Micromechanical modeling of the microbond test to quantify the interfacial properties of fiber-reinforced composites. International Journal of Solids and Structures July, 2008, 45(14–15), 4098–4113.

12. Sockalingam, S. Fiber-Matrix Interface Characterization through the Microbond Test, Int'l Journal of Aeronautical and Space Science [Online] 2012, 13(3), 282–295 http://ijass.org/On_line/admin/files/(282–295)Review%EC%B9%BC%EB%9D%BC.pdf (accessed May 2, 2016).

13. Loan, T. T. Investigation on jute fibers and their composites based on polypropylene and epoxy matrices, PhD Theses, Der Fakultät Maschinenwesen der Technischen Universität Dresden, 2006.

14. MATERIAL Type Cost ($/kg) Density (ρ, mg/m3) Young's Modulus. http://ocw. mit.edu/courses/materials-science-and-engineering/3-11-mechanics-of-materials-fall-1999/modules/props.pdf (accessed May 2, 2016).

15. Shaha, D. U.; Schubela, P. J.; Licenceb, P.; Clifforda, M. J. Determining the minimum, critical and maximum fiber content for twisted yarn reinforced plant fiber composites. Compos. Sci. Technol. October, 2012, 72(15), 1909–1917.

16. Al-Bahadly, E. A. O. The mechanical properties of natural fiber composites, PhD Theses, Faculty of Engineering Swinburne University of Technology in January, 2013.

17. Ihuezea, C. C.; Okafora, C. E.; Okoye, C. Natural fiber composite design and characterization for limit stress prediction in multiaxial stress state, Journal of King Saud University – Engineering Sciences, July, 2015, 27(2), 193–206.

18. Halpin, J. C.; Kardos, J. L. The Halpin–Tsai equations: a review. Polym Eng Sci. 1976, 16(5), 344–52.

19. Facca, A. G.; Kortschot, M. T.; Yan, N. Predicting the elastic modulus of natural fiber reinforced thermoplastics, Composites: Part A 2006, 37, 1660–1671.

20. Lu, Y. Mechanical Properties of Random Discontinuous Fiber Composites Manufactured from Wetlay Process, Thesis, Faculty of the Virginia Polytechnic Institute and State University, 2002.

21. Chen, P. E. Strength Properties of Discontinuous Fiber Composites," Polymer Engineering and Science 1971, 11(1), 51–56.

22. Venkateshwaran, N.; Elaya Perumal, A. Modeling and evaluation of tensile properties of randomly oriented banana/epoxy composite, JRPC, Published online before print December 1, 2011, 1–11 (accessed May 13, 2016).

23. Hickie T.; Chaikin M. Some Aspects of Worsted-Yarn Structure Part III: The Fiber-Packing Density in the Cross-Section of Some Worsted Yarns. J. Text. Inst., 1973, 433–437.

24. Saleh, M.; Johari M.; Amani M. The Effect Of Process Variables On The Packing Density Of Local Ring Spun Yarns, Autex 2009 World Textile Conference 26–28 May, 2009 Izmir, Turkey.
25. Driscoll, R.; Postle, R. Modeling the Distribution of Fibers in a Yarn. J. Text. Inst., 1988, 79(1), 140–143.
26. Pan, N. Theoretical determination of the optimal fiber volume fraction and fiber–matrix property compatibility of short fiber composites. Polym. Compos. 1993, 14(2), 85–93.
27. Cann, M. T.; Adams, D. O. Characterization of Fiber Volume Fraction Gradients in Composite Laminates. J. Compos. Mater. March, 2008, 42(5), 447–466.
28. Wasik, T. Effect of fiber volume fraction on fracture mechanics in continuously reinforced fiber composite materials, PhD Theses, University of South Florida, USA, 2005.
29. Zimniewska, M.; Radwanski, M. Natural Fiber Textile Structures Suitable for Composite Materials. J.NAT. FIBERS 2012, 9(4), 225–239.
30. Hickie, T.; Chaikin, M. Some aspects of worsted-yarn structure Part III: the fiber-packing density in the cross-section of some worsted yarns. J. Text. Inst. 1973, 433–437.
31. Kremenakova, D. Methods for investigation of yarn structure and properties. [Online] http://centrum.tul.cz/centrum/itsapt/Summer 2004/files/kremenakova1.pdf (accessed Dec 15, 2015).
32. Salehi, M.; Johari, M.; Amani, M. The effect of process variables on the packing density of local ring spun yarns, Autex 2009 World Textile Conference, Izmir, Turkey, 26–28 May, 2009.
33. Hussain, U.; Sarwar, A.; Shafqat, A. R.; Iqbal, M.; Zahra, N.; Ahmad, F.; Hussain, T. Effect of spinning variables on yarn packing density. Indian Journal of Fiber & Textile Research. December, 2014, 39, 434–436.
34. Morris, P.; Merkin, J.; Renneh, R. Muzaffar, M.; Modeling of yarn properties from fiber properties, J. Tex. Inst. Part I. 1999, 3, 322–335.
35. Yilmaz, D.; Goktepe, F.; Goktepe, O.; Kremenakova, D. Packing density of compact yarns, Text. Res. J. 2007, 77, 661–667.
36. El Messiry, M. Theoretical analysis of natural fiber volume fraction of reinforced composites, Alexandria Engineering Journal 2013, 52, 301–306.
37. Goutianosa, S.; Peijsb, T.; Nystromc, B.; Skrifvarsd, M. Textile reinforcements based on aligned flax fibers for structural composites. [Online] http://citeseerx.ist.psu.edu/viewdoc/download?doi=10.1.1.470.1201&rep=rep1&type=pdf (accessed May 7, 2016).
38. Reiter Rikipddia. http://www.rieter.com/En/Rikipedia/Articles/Technology-Ofshort-Staple-Spinning/Yarn-Formation/Assembly-Of-Fibers-To-Make-Up-A-Yarn/Number-Of-Fibers-In-The-Yarn-Cross-Section (accessed May 7, 2016).
39. Lehtiniemi, P.; Kari Dufva, K.; Berg, T.; Mikael Skrifvars, M.; Jarvela, P. Natural fiber-based reinforcements in epoxy composites processed by filament winding. J. Reinf. Plast. Compos. 2011, 30(23), 1947–1955.
40. Gibson L.; Batch, C. S.; Christopher W.; Compaction of Fiber Reinforcements. Polym. Compos. June, 2002, 23(3), 307–318.

41. El Messiry, M. Design of multilayer filters, study the effect of nanofiber structures application, Autex 2011 World Textile Conference, Mulhouse, France, June 8–10, 2011.

42. Berkalp, O. Air permeability & porosity in spun-laced fabrics, fibers & textiles in Eastern Europe, July/September 2006, vol. 14, No. 3(57), 81–85.

43. Havlova, M. Influence of vertical porosity on woven fabric air permeability, in: 7th International Conference Texsci. 2010, September 6–8, Liberec, Czech Republic.

44. Turan, R.; Okur, A. Investigation of pore parameters of woven fabrics by theoretical and image analysis methods. J. Tex. Inst. 2012, 8, 875–884.

45. Ihueze, C. C.; Okafor, C. E.; Okoye C. I. Natural fiber composite design and characterization for limit stress prediction in multiaxial stress state, Journal of King Saud University – Engineering Sciences 2015, 27, 193–206.

46. NASA TECHNICAL STANDARD, Structural design and test factors of safety for spaceflight hardware, NASA technical standard, NASA-STD-5001B. [Online] 2014 https://standards.nasa.gov/file/506/download?token=9r45Bxy4 (accessed Dec. 2, 2015).

47. Guide to Verifying Safety-Critical Structures for Reusable Launch and Reentry Vehicles Version 1.0, Federal Aviation Administration. [Online] 2005 https://www.faa.gov/.../RLV_Safety_Crit... (accessed Dec. 2, 2015).

48. National Aeronautics and Space Administration, Structural Design Requirements and Factors of Safety for Spaceflight Hardware for Human Spaceflight October 21, 2011. [Online] http://ntrs.nasa.gov/archive/nasa/casi.ntrs.nasa.gov/20110023499.pdf (accessed Dec. 31, 2015).

49. Echtermeyer, A. T.; Lasn, K. Safety factors and test methods for composite pressure vessels, FCH-JU-2009–1 Grant Agreement Number 278796. [Online] 2013 http://www.fch.europa.eu/sites/default/files/project_results_and_deliverables/Report%20on%20safety%20(ID%202849589).pdf (accessed May 5, 2016).

50. Nkiwane, L. Structure, creep and fatigue relationships of nylon 6.6 tire materials, ZJST. [Online] 2010, 5, 46–54, http://ir.nust.ac.zw/xmlui/ (accessed May 5, 2016).

51. Mahadevan, S.; Mao, H. Probabilistic Fatigue–Creep Life Prediction of Composites, Journal of Reinforced Plastics and Composites. March, 2004, 23(4), 361–371.

52. Mao H., Mahadevan S.; Ghiocel D., Probabilistic Creep-Fatigue Life Prediction of Aircraft Engine Blades. 15th Engineering Mechanics Conference, ASCE, 2002, New York, NY, USA.

53. Zhuang, W. Z.; Swansson, N. S. Thermo-Mechanical Fatigue Life Prediction: A Critical Review, Airframes and Engines Division Aeronautical and Maritime Research Laboratory. [Online] http://www.ewp.rpi.edu/hartford/~ernesto/F2011/EP/MaterialsforStudents/DeRosa/Zhuang1998-DSTO-TR-0609%20PR.pdf (accessed May 6, 2016).

PART III

COMPOSITE MANUFACTURING TECHNIQUES AND AGRICULTURE WASTE MANUFACTURING

CHAPTER 6

NATURAL FIBER COMPOSITES MANUFACTURING TECHNIQUES

CONTENTS

6.1   INTRODUCTION

The manufacturing of natural fiber composite (NFPC) includes the process of interfere the matrix into the reinforcement textile structure material to produce the final composite shapes. The reinforcement may be in form of wood flour, short fibers in bulk form (two-dimensional fiber mat or random fiber mat), entangled forms (non-woven), yarns, 2-D or 3-D fabrics (woven, knitted, biaxial braided, triaxial, multidirectional). The manufacturing process should satisfy the following requirements:

1.   Not affect the orientation of fibers or yarns of the reinforcement architect;
2.   The fiber volume fraction should be constant in the composite in its different parts;
3.   Insure excellent bonding of the matrix to the fiber reinforcement;

4. Not allow yarns touch each other;
5. Insure minimum voids percentage;
6. Minimum material waste of fibers or matrix;
7. Low cost;
8. Safety and environment protection.

Consequently, the choice of the manufacturing method has a significant sequence on the performance of the composite through the anisotropic behavior reduction of the composite and delamination, to insure the maximum utility of the fiber strength as main target of the composite design. Environment protection issues should be addressed, too. The U.S. Environmental Protection Agency has continued to strengthen its requirements to meet the mandates of the Clean Air Act Amendments, passed by Congress in 1990 [1]. Specifically, the agency's goal is to reduce the emission of hazardous air pollutants (HAPs), a list of approximately 180 volatile chemicals that are considered of health risks. The EPA enacted regulations specifically for the composites industry, requiring emission controls using maximum achievable control technology (MACT) standards [2].

6.2 MANUFACTURING OF COMPOSITES

Various methods are existed for the manufacturing of the composite material, such as:
1. Extrusion
2. Injection molding
3. Structural foam molding
4. Rotational molding
5. Thermoforming
6. Compression molding
7. Casting processing of reinforced plastic
8. Open mold processing.

The following are examples, given in Table 6.1, of the technologies used in the production of various types of natural fibers-based composites [3–5]. The production technique is influenced by the maximum fiber volume fraction of the final composite.

The evaluation of the different methods of composite manufacturing and their applicability to be used in the NFPC are given in Table 6.2.

TABLE 6.1 Manufacturing Methods, Materials, Products of Fiber Polymer Composite [3–5]

Product	Reinforcement	Matrix	Technology	Applications	Fiber volume fraction
Fiber boards	Non-woven fabrics	Thermosetting	Molding under hydraulic pressing	Flooring, roof, internal walls and panels for automotive and furniture industry	50–70%
3-D shaped fiber boards	Natural fiber non-woven mats and granulated natural fibers and wood flour	Polyester and thermoplastic high density or low density polyethylene	Extrusion molding, compression molding	Automotive interiors, furniture	40–50%
Medium Density Fiberboard (MDF)	Bagasse mixed with other agricultural fibers	Thermosetting resin, Phenol formaldehyde resin, Methylene diphenyl diisocyanate, Urea-formaldehyde	Fiber de-fiberized for mat formation	Solid wood replacement	50–70%
Pultrusion profiles	Natural fibers (jute, sisal, ramie) in mats, fabrics or hybrids	Thermosetting liquid resin: Polyurethanes, Epoxy resin, Polyimides, Polyester resin, Urea-formaldehyde	Pultrusion	Different profiles	50–70%
Long Fiber Reinforced Thermoplastic	Natural fibers (sisal, jute, etc.) in roving or yarns forms	Thermoplastic polymer: Polypropylene, polyethylene, ethylene propylene rubber	Extrusions with feeders of yarns	Different profiles as a replacement for solid wood	40–50%

TABLE 6.1 (Continued)

Product	Reinforcement	Matrix	Technology	Applications	Fiber volume fraction
Molded products	Nonwoven mat	Unsaturated polyester resin	Resin Transfer Molding	Door siding, components and parts for automotive industry, instrument panels, engine covers, etc.	30–40%
Granulate of natural fibers blended with thermoplastic resin	Natural granulated fibers	Thermoplastics resin	Melting and composting in twin-screw extruder	Pallets, packages, appliances	40–50%
Thermoplastic residue boards	Agriculture residues (straw, bark, fiber fines, bagasse, etc.).	Thermoplastics	Extrusion (single-screw) and compression molding	Exterior furniture wall panels, wire electric coils, etc.	40–50%
Hybrids of natural and glass fibers	Natural fibers/glass fibers	Unsaturated polyester and epoxy resin.	Resin Transfer Molding, vacuum injection	Boats, water containers, storage grains, etc.	30–40%
Roof shingles	Natural fibers bundles, Non-woven mats	Cement and blast furnace slag mortar	Concrete manufacturing technology	Civil engineering replacement of asbestos fibers and concrete, etc.	According to end use
Cellular concrete	Rice, sisal, jute, sugarcane bagasse, ramie residues	Clay	Concrete manufacturing technology	Interiors in high buildings for weight reduction	According to end use

TABLE 6.2 Manufacturing System Evaluation

System evaluation regarding the manufacturing of NFPC	Applicability NFPC	Versatility	Waste	Automation	Heat requirement	Environmental hazard*	Low cost
Open molding (Spray fiber and polymer)	++++	+++	+++	+++	+	++++	++++
Resin transfer molding (RTM)*	++++	+++++	+++++	++++	+	++	++++
Resin injection molding (RIM)	+++++	+++++	+++++	+++++	+	++	++++
Vacuum-assisted resin transfer molding (VARTM)	+++++	+++++	+++++	+++++	+	++	++++
Resin film infusion (RFI)	++++	+++	+++	+++	++++	++	+++
Compression molding (SMC)	+++	+++++	+++++	+++++	++++	++	+++++
Injection molding	+++++	+++++	++++	+++++		+	
Filament winding (thermoplastics)	++++	+++	++++	+++++	+++++		++++
Centrifugal casting (thermoplastic)	+++	+	+++	+++++	+++	++	++++
Extrusion	+++++	+++	+	+++++	+++	++	+++++
Hand layup	++++	++	++++	−	++++	++	++
Pultrusion	+++	+++	+	+++++	+++	++	++++

*Environmental hazard: hazardous air pollutants, + sign means high and − sign means low.

6.3   METHODS OF MANUFACTURING COMPOSITE MATERIALS

Fiber reinforced polymer composites are made of polymer resins as the matrix and different types of fibers as the reinforcement. The demand for advanced composite materials has been on the rise in fields as automobile manufacturing, aerospace, civil constructions, and shipbuilding. Several manufacturing techniques may be used for the production of natural fiber composites (NFPC). They differ according to the type of resin used, either thermoset or thermoplastic resins, as well as the cost, requirements for making net-shaped parts, and cure processes temperature. Several studies indicate the inter-relation between composite manufacturing and final composite properties, composite defects, cost, and their performance [5–16]. For natural fiber composites the following manufacturing methods are recommended:

1.   **For thermosetting composites**: hand lay, spray, resin transfer molding, compression injection, pressure bag, vacuum assistance, Pultrusion.

2.   **For thermoplastic composites:** extrusion, injection molding, compression, cold press, filament winding and rotational molding.

6.3.1   PULTRUSION

The fiber reinforcement may be in the form of continuous yarns wound on packages or fabric or both types. The required number of the yarns to form the reinforcement are fixed on creel and arranged to form a horizontal sheet of yarns at the entrance of the machine. The yarn sheet is pulled through the polymer in Pultrusion tank, where the polymer will be infused into the yarn sheet that passes through the preform die which defines the thickness and width of the composite material. The preform will be subjected to heat in the heated die in order to cure the composite. The composite material will move using a transporting belt and wound at the end of the machine. In multi-layer laminate, fabric can be fed on special creel and passed through the same parts of the machine with the yarn sheet, as shown in Figure 6.1. The pultruded fabric may be rolled or lay in form of sheet. There are various designs to the Pultrusion machines which

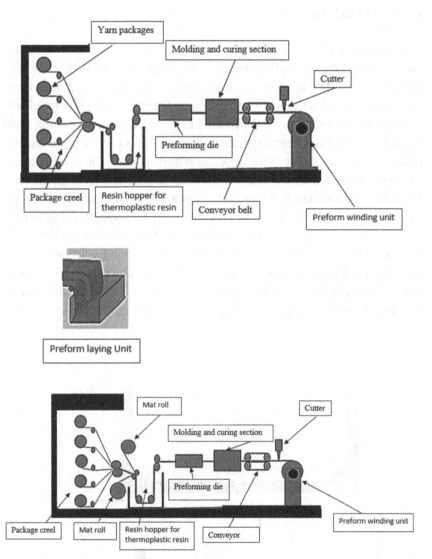

FIGURE 6.1 Principles of Pultrusion process.

are designed to deal with certain composite materials and products. The degree of automation on such machines makes it appropriate for the production of laminate composites. The feeding section has several designs to suite the required composite structure.

6.3.2 HAND LAYUP

This is the simplest and widely used for many years way for the composite formation, especially for large size composites, such as in the boat manufacturing. The reinforcement in this case will be in the form of mat of woven, knitted, or nonwoven. The short fibers or yarns require efforts to get uniform thickness of the formed preform. The thermosetting resin is used and the impregnation of the reinforcement is carried out manually using roller brush to distribute the resin on the surface of the reinforcement and apply enough pressure to allow the resin to penetrate through the reinforcement without voids, Figure 6.2. The preform may consist of several laminates, so the process will be repeated layer by layer. Resin used may be epoxy, polyester, vinyl ester, and phenolic. However, the quality of the preform depends on the skills of the labor. Due to the high fiber volume ratio, the probability of void formation increases. In order to reduce these risks, a vacuum bag is used to suck the air from the unreachable spaces, helping the resin to reach them. Hand lay or open mold technique is also

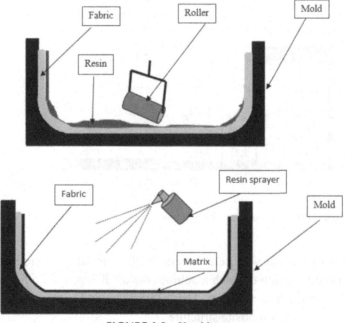

FIGURE 6.2   Hand layup.

connected with feeding of the fibers which are cut into small length, may reach micro size, mixed with the polymer and sprayed in the open mold.

6.3.3 FILAMENT WINDING

In this process the yarns are cross-wound on a rotating mandrill to form the final form of the reinforcement. The fiber yarns are passing through a resin bath in order to impregnate the yarn with resin before it wound over the mandrel, as shown in Figure 6.3. The resin may be of any type that

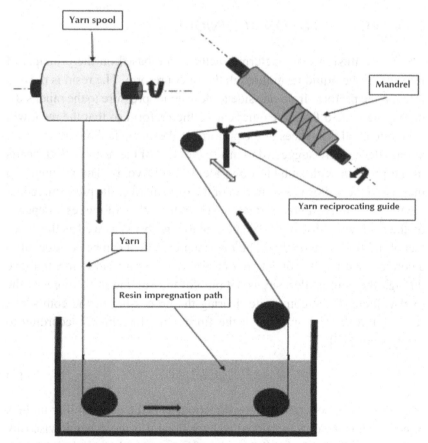

FIGURE 6.3   Sketch of the filament winding.

satisfies the designed requirements, i.e. epoxy, polyester, vinylester, phenolic. The percentage of the resin take-up can be controlled by adjusting
the winding speed and the resin viscosity. The structure of the preform is
very uniform and the fiber volume fraction is uniform all over the composite. The mandrel can be a part of the design or can be removed after
the formation of composite shape through special mandrel disengaged
design. This technique of composite manufacturing is of reasonably low
cost but limited to 3-D curved shapes. After preparation of the composite
with the required thickness, it is cured and the mandrel removed. Composite using such method of preparation is suitable for the parts subjected
to pressure.

6.3.4   RESIN INFUSION TECHNIQUE

The resin infusion manufacturing methods are based on the principle of
infusion of the liquid resin through the fiber preform. The resin is pressed
through the preform from one side to flow under pressure to the other side,
or may be sucked from the other end of the preform so that the resin will
penetrate in all the spaces of the preform as the air sucked out. In practice,
several designs are suggested to insure the flow of the resin in all the parts
of the preform with suitable flow rate, to avoid voids. The vacuum bag
may assist the uniform distribution of the resin all over the preform, reducing the possibility of void formation. The rate of flow of the resin depends
on the resin viscosity, fiber volume fraction, porosity, as well as the architect of the fibrous assembly. The fiber reinforcement can be considered as
a porous material. Its porosity defines in what way a fluid is able to move
through the porous. Porosity is the ratio of the volume of the pores to the
total volume of the composite. During the processing of the composite,
resin is forced to flow through the fibrous reinforcement, according to
Darcy's law [17]:

$$q_x = (k_x \, A/\mu) \, (\delta P/\delta x) \tag{1}$$

where: q_x is the flow rate in the x direction; k_x is the permeability in the x
direction; A is the cross sectional area for flow; μ is the dynamic viscosity
of the resin; $(\delta P/\delta x)$ is the differential of pressure in the x direction.

The permeability coefficient k depends on the combination of the polymer and porous reinforcement used. The greater the value of k, the higher will be the rate of flow of a fluid through a material. The resin flow rate through the reinforcement is a function of differential pressure, the inversely proportional to the resin dynamical viscosity and reinforcement porosity. In the case of natural fiber composite, the situation is more challenging, hence an increase in pressure will increase the velocity of the fluid though the enforcement. However, increasing the pressure of the fluid will increase the compaction pressure and lower the permeability. For instance fabric compaction also impacts the porosity of the fabric, which will affect the saturation time. Although permeability decreases with compaction, the decrease in porosity can increase the velocity of the fluid through a preform. Decreasing the porosity also increases the capillary pressure. Also there is another category of resin flow, micro-flow which is considered to be the flow inside the fiber yarns and macro-flow is considered to be the flow in the macro-pores between yarns [18]. The pressure needed to move the resin through the matrix should be enough to develop a flow of the resin through the yarns and between the yarns and cause no fiber or yarn compaction, depending on the permeability and the fluid viscosity [19]. Good composite is achieved when saturated fabric reaches that all the pores are filled with the resin. The fabric porosity and resin viscosity are the key parameters that govern the resin flow in the fiber reinforcement, especially for those manufactured under low pressure. This problem has been studied by several investigators [20–29]. When a fluid travels through a dry reinforcement of natural fibers, different mechanisms occur. For instance, in woven fibers preforms, a macroscopic flow through the inter-yarn regions takes place, followed by a delayed microscopic flow through the intra- yarn region. The fabric porosity depends on the fabric design: the number of intersections, the yarn count, weft and warps per cm, crimp percentage, the yarn twist per cm, and fiber packing density, consequently the value of k which depends on the fabric porosity, it increases with the increase of fabric porosity [29, 30]. For that reason the estimation of pores in the fabric has been intensively investigated [31–37]. Analysis of the type of pores, which are void spaces within the material and are separated from each other, are classified according to position in the material [37] into inter-pores or intra-pores. The pore width can be classified as pore-width b>50 nm, b<50 and >2 nm or b<2 nm. Not all the pores are accessible to the flow of the resin, open pores connected to the outer surfaces permit fluid flow, while "blind pores" are on the contrary. In connection with nature fibers reinforcement, all the types of the pores are expected. In the fabric laminates, due to the increase of fabric micro porosity, resin flow becomes more important [38] to be considered. For fiber mat where the fibers randomly oriented the situation will be more dependent on the mat packing density, in some cases high pressure may be needed to fill the intra-pores.

6.3.4.1   Liquid Composite Molding (LCM)

Liquid Composite Molding is a technology to manufacture net-shaped composite parts mod with a thermosetting resin. First, a preform is created from reinforcing fibers. Next, the preform is inserted in a mold that matches the dimensions of the desired part. The mold may be solid mold or soft mold. Low viscosity thermosetting resin, mixed with a hardener, is injected under pressure. The resulting part is cured at room temperature or under a strictly controlled mold-temperature cycle till the end of the curing process (Figure 6.4). In some cases, the mold is covered by a flexible sheet and the resin is applied under vacuum assistance to prevent the void's formation, so that high fiber volume fraction can be obtained with very low void contents. The viscosity of the resin as well as the pressure applied during the infusion is very essential to get composite parts voids free. Small complex automotive components and seats can be manufactured using this technique. The resin transfer in the RTM is assisted by the compaction of fibrous preform that will change the alignment of the fibers in the molding process.

6.3.4.2   Resin Film Infusion (RFI)

For the reduction in the resin cost in the composite formation, resin in the form of film is used and laid under dry reinforcement in the mold which is heated to reduce the viscosity of the resin film, as shown in Figure 6.5. The mold is heated in oven for melting the resin film, then followed by the cure

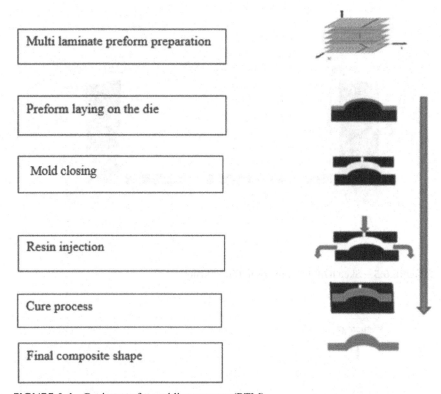

FIGURE 6.4 Resin transfer molding process (RTM).

cycle. The use of vacuum bag results in that the resin is defused through the reinforcement. This process seeks high equipment cost.

6.3.4.3 Resin Vacuum Infusion (RVI)

The Vacuum Infusion Process (VIP) is a technique that uses vacuum pressure to drive resin into a laminate. Figure 6.6 illustrates vacuum infusion principles. Materials are laid dry into the mold and the vacuum is applied before resin is introduced. Once a complete vacuum is achieved, resin is literally sucked into the laminate via carefully placed tubing. The resin tap, Figure 6.6, is connected to the resin reservoir. Vacuum Infusion Processing is usually using vacuum bag for more control of the resin flow inside the preform, allowing increased laminate compression, a higher fiber-to-

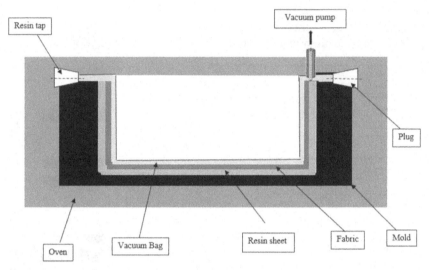

FIGURE 6.5   Sketch of the of the resin film infusion.

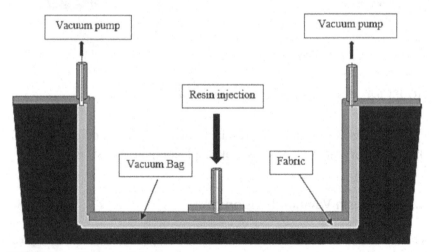

FIGURE 6.6   Vacuum infusion.

resin ratio, and good specific strength characteristics. The advantages of vacuum infusion are: less percentage of voids; less wasted resin and very consistent resin usage; and besides it is environmentally friendly process [39]. RVI process can be applied to manufacture of large parts utilizing low viscosity resin.

6.3.5 3-D PRINTING FOR NATURAL FIBER POLYMER COMPOSITE

Using the modern technology, the natural fiber can be reduced in size to get Nano and micro particles. In this form it can be mixed with polymer using current techniques for the formation a complicated shapes from NFPC, such as 3-D printing (first commercial machine 1988). This technology allows a great freedom and fast design to produce new products with less material waste. Moreover, the processing cost is reduced [40–47]. The general principles of 3-D printing are illustrated in Figure 6.7, which are:

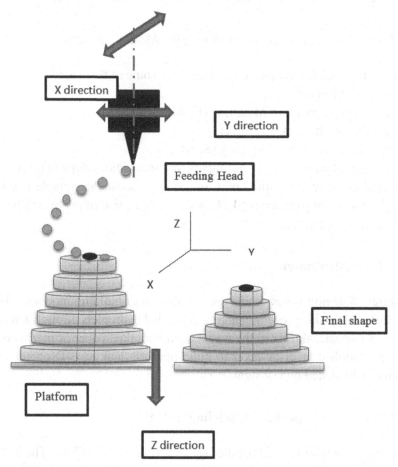

FIGURE 6.7 Schematics of building a tubular composite on 3-D printer.

Step I: 3-D scanning of the design or using computer aided design software program

Step II: Feeding the file to the printer

Step III: The computer will slice the part image with distance Δh varied between 0.05 and 0.15

Step IV: The printer will deposit polymeric material on the platform first layer

Step V: The platform will move down the designed layer thickness Δh, the smaller its value – the accurate shape of the final designed part

Step VI: UV or Laser are used to harden the deposit layer.

6.3.5.1   Techniques of 3-D Printing for Model Formations

There are several principles of 3-D printers, some of them are:

1.   3-D Printer
2.   Fuse Deposition Modeling (FDM)
3.   Stereolithography (SLA)
4.   Selective Laser Sintering (SLS)

Each method is found to be suitable for applications depending on the material of the manufactured part. In order to choose the suitable one for the processing of parts from NFPC, a short description of the principles of 3-D printers are given.

6.3.5.2   3-D Printer

Figure 6.8 demonstrates schematics of method of building model on 3-D printer [43]. In this process, material is applied in droplets through a small diameter nozzle, similar to the way a common inkjet paper printer works, but it is applied layer-by-layer to build platform making 3-D object and then it is hardened by UV light.

6.3.5.3   Fuse Deposition Modeling (FDM)

The most commonly used technology in this process is FDM. The FDM technology works using a plastic filament or metal wire which is unwound

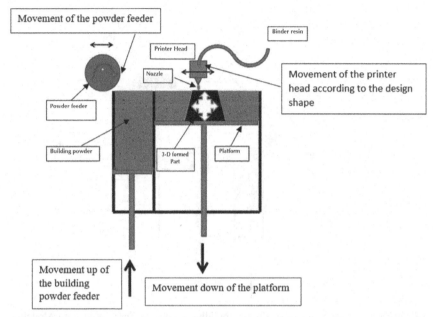

FIGURE 6.8 Schematics of method of building model on 3-D printer.

from a coil and supplying material to an extrusion nozzle which can turn the flow on and off. The nozzle is heated to melt the material and can be moved in both horizontal and vertical directions by a numerically controlled mechanism, directly controlled by a computer-aided manufacturing (CAM) software package. As the nozzle is moved over the table in the required geometry, it deposits a thin bead of extruded plastic to form each layer. The plastic hardens immediately after being squirted from the nozzle and bonds to the layer below. Figure 6.9 shows the schematics of building model on FDM [43].

6.3.5.4 Selective Laser Sintering (SLS)

This technology uses a high power laser to fuse small particles of plastic, metal, ceramic or glass powders into a mass that has the desired three-dimensional shapes. The laser selectively fuses the powdered material by scanning the cross-sections (or layers) generated by the 3-D modeling

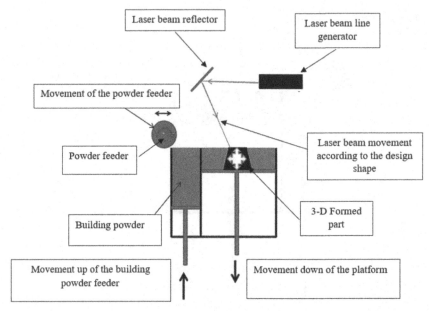

FIGURE 6.9 Schematics of method of building model on Fuse Deposition Modeling (FDM).

program on the surface of a powder bed. After each cross-section is scanned, the powder bed is lowered by one layer thickness. Then a new layer of material is applied on top and the process is repeated until the object is completed. The process is repeated until the whole model is completed.

The build platform is raised and the loose powder is vacuumed away, revealing the completed part. Figure 6.10 illustrates the principles of SLS methods [43].

6.3.5.5 Stereolithography (SLA)

The most commonly used technology in these processes is SLA. This method was the first RP method utilized by the industry. The method employs a ultra-violet laser beam to cure liquid photo polymer in layers. A platform starts one layer depth below the surface of the liquid, and the laser cures the first layer. The edge of the layer is drawn out and then the

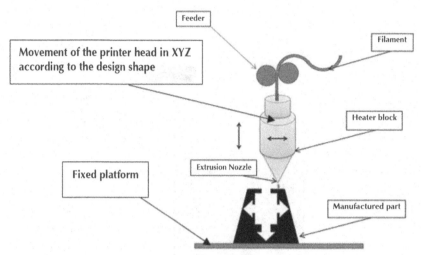

Feeder

Filament

Movement of the printer head in XYZ according to the design shape

Heater block

Fixed platform

Extrusion Nozzle

Manufactured part

FIGURE 6.10 Schematics of method of building model on Selective Laser Sintering (SLS).

interior is hatched to solidify inside the object. Then the platform is lowered by one layer depth and the next layer is cured. The process is repeating until the object is completed. Figure 6.11 illustrates the principles of SLA [43].

FDM principle seems to be the most suitable for the processing of natural fiber particles composites. The 3-D printing technology can significantly improve the technology of composite manufacturing. Hence, it can give fast and reliable products of tailored-made parts. Natural fiber reinforced thermoplastics are of high-performance as Bio composite materials [48]. Spherical pellets can be processed on 3-D printer using FDM.

In the last decade, filament which is a mix of recycled wood fibers and polymer binder, can be melted and extruded just like any other on 3-D printer [45, 49]. One interesting feature of this material is the ability to add 'tree rings,' or a subtle gradation in color from a rich brown to very nice beige as illustrated in Figure 6.12. It is expected that the use of the wood flour or wood microfiber will give way to the new products on 3-D printing.

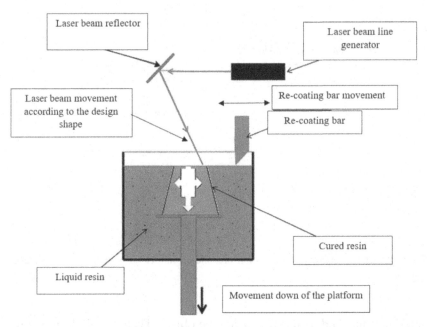

FIGURE 6.11   Schematics of method of building model on Stereolithography (SLA).

FIGURE 6.12   Wood composite produced on 3-D printing.

6.3.6   EXTRUSION

This process is common for thermoplastic material. It provides the advantages of high production rates and possible products with a good finish and a good consistent quality. Wood husk and straw can be processed with polyethylene (PP), polypropylene (PE) and polyvinylchloride (PVC) for the production of WPC using extrusion technique. Wood flour is mainly produced from the residues of various wood processes, and subjected to milling processes for size reduction. Wood plastic composites (WPC) made out of wood waste and plastic waste has been a fast growing research area because of its wide range of applications [50–53]. The manufacturing of wood-plastic composites starts with wood material preparation, blending with polymer, and molding. Various techniques were suggested for molding, basically injection molding, profile extrusion or compression molding. There are also wood pellets which provide pre-processed wood fibers as fillers for many applications. The fiber volume fraction may reach 30–40%.

Molten polymer goes through a die to produce a final shape. It involves four steps:

- Wood flour with different particle sizes, 74–500 μm [54] blended by the required ratios;
- Wood flour, dried for 3–4 hr. at 105 to 115°C;
- Wood pellets and the polymer are mixed with coloring and additives (coupling agents, light stabilizers, pigments, lubricants);
- The granular plastic is blended with dry wood flour and fed in a hopper of extruder, as illustrated in Figure 6.13;
- Extruder screw thread turns forcing the wood-plastic material through a heater, melting it;
- The extrusion process is carried out at the temperature ranging 140°C÷185°C that corresponds to the temperature of plastic granulates preparation and increases in the direction of the material flow;
- The wood-plastic material is forced through a die to produce the required shape;
- Cooling process is followed which continued for predetermine time.

Several types of polymers can be used in the extrusion processes, such as Polyethylene Terephthalate (PET or PETE), High density Polyethylene (HDPE), Polyvinyl Chloride or Vinyl (PVC-V), Low density Polyethylene (LDPE), Polypropylene (PP), Polystyrene (PS) and all have a potential for recycling.

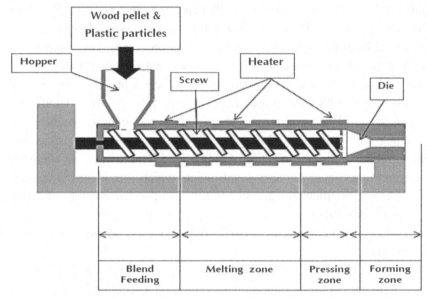

FIGURE 6.13   Principles of single screw extruder.

6.4   COMPOSITE DEFECTS

The composite under actual loading may be subjected to one or all the following defects, fraction of fibers, failure of the interface between fiber and matrix, delamination, debonding failures, and cracks. These modes of failures are due to the defects either in the fiber, fabrics, polymer, or by manufacturing technology. In the modern industry as production rates are expected to expand, the quality target must be zero defects, zero rework and repair and zero scrap [55]. Consequently, analysis of the composite defects in each case become the essential purpose of the quality control of the composite production. The main sources that causing defects in the composite materials are as illustrated in Figure 6.14.

The types of defects can be classified according to their sources as given in Figures 6.15–6.19.

The analysis of the sources of defects [56–60] in natural fiber polymer composites are given in Figure 6.16, while the other figures give the analysis of the defect's sources for each component.

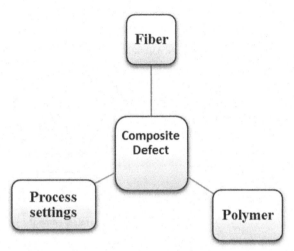

FIGURE 6.14 Sources of composite material defects.

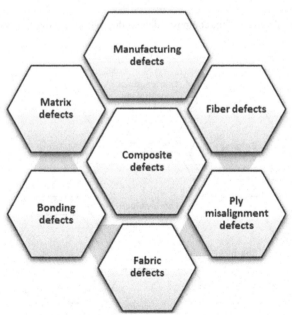

FIGURE 6.15 Analysis of the composite defects.

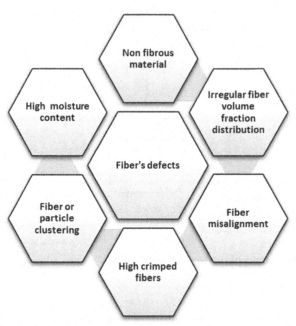

FIGURE 6.16   Analysis of the composite defects due to fibers, yarns or particle properties.

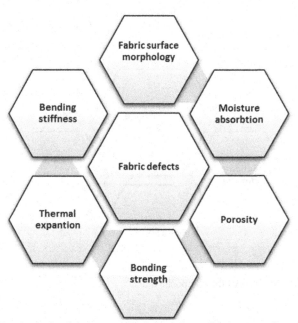

FIGURE 6.17   Analysis of the composite defects due to fabric properties.

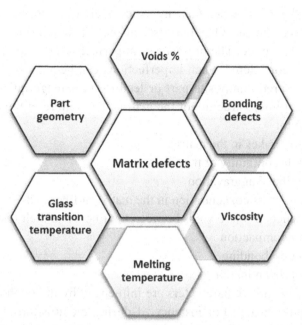

FIGURE 6.18 Analysis of the composite defects due to matrix properties.

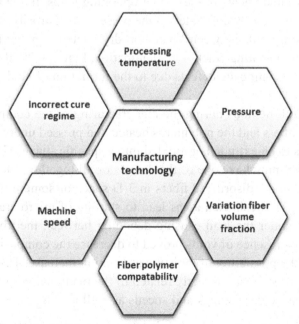

FIGURE 6.19 Analysis of the composite defects due to manufacturing process.

The final performance of a composite depends not only on the choice of the matrix and the fiber but also on the manufacturing process by which they are made. The presence of imperfections due to manufacturing must be considered. Such imperfections can be already damage for the manufactured composite part or lead to its damage quickly due to the change of microstructure of the matrix. The failure of the composite may be due to:

1. Micro crakes in the matrix
2. The discontinuity of the fibers
3. The fibers' aggravation
4. Shear stress concentration in the matrix at fiber ends
5. Presence of micro voids between the fibers and their distribution
6. Fiber compaction
7. Fiber de-bonding
8. Fiber delamination

Most of the above parameters are influenced by the method of composite manufacturing. For instance, if during manufacturing the fibers are distributed unevenly in the volume of the composite, this will generate intra-laminar shear stresses under operating loads. In the matrix, the damage is essentially correlated to the presence of porosity. The formation of micro voids between fibers and dry spots are potential starting points for propagating cracks or delamination. Figure 6.20 illustrates the causes of the composite defects due to the formation of voids and micro cracks.

In several methods during the manufacturing of the composites, the mix of the fibers and the polymer is heated and pressed under high-pressure values to be extruded or molded into a specific shape. The resultant cross section may have one of the defects or all together, such as resin-rich areas, voids, disordered fibers in 3-D space or sometimes the fiber may break. The resin-rich areas lead to starting of micro cracks which will speed under the load causing deboning between the fiber and the matrix. The presence of voids proved to decrease the composite strength as well as the possibility of moisture absorption increase. The defects in the final composite are directly related to the manufacturing techniques and their particular settings and speeds as well as the shape of the processed parts.

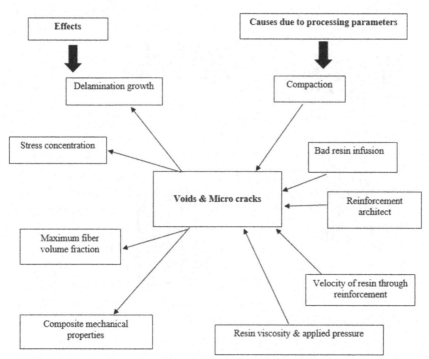

FIGURE 6.20 Manufacturing parameters and material behavior effects on voids and micro crack formation.

The analyses given in Figure 6.20 are the general overview of the composite defects. Figure 6.21 illuminates some of these defects, which are due to composite materials behavior effects of voids and micro crack and manufacturing.

Referring to the interlayer delamination, one of the laminates will be separated from the other laminate, causing the micro cracks, vice versa to the outer layer delamination. The delamination may be on one place or different portions of the matrix. There is deference between deboning and delamination – the later means the laminates become separated while deboning happens when matrix and fiber doesn't adhere together. Incorrect cure occurs owing to the incorrect formulation or mixing, or thermal exposure conditions of the fibers or matrix.

Non-destructive testing (NDT) for defects should be minimized by achieving manufacturing processes which are under control.

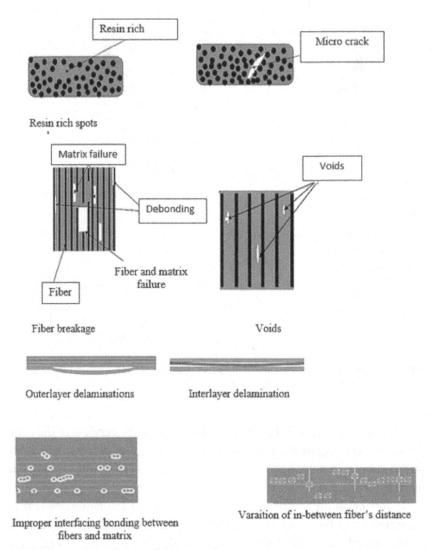

FIGURE 6.21 Some defects of the composites.

KEYWORDS

- **3-D printer**
- **composite defects**
- **compression molding**
- **extrusion**
- **hand layup**
- **injection molding**
- **manufacturing of composites**
- **open molding**
- **pultrusion**
- **resin film infusion**
- **resin injection molding**
- **resin transfer molding**

REFERENCES

1. EPA. Clean Air Act – The Clean Air Act Amendments of 1990 [Online] https://www.epa.gov/clean-air-act-overview/1990-clean-air-act-amendment-summary (accessed May 7, 2016).
2. NESHAPS – Maximum Achievable Control Technology (MACT) Standards. The regulations took effect in early 2006 [Online]. http://www.epa.gov/wastes/hazard/tsd/td/combust/index.htm (accessed May 7, 2016).
3. Trade and Markets Division, Unlocking the commercial potential of natural fibers, market and policy analysis of non-basic food agricultural commodities team. FAO of the UN publications [Online] 2012. http://www.fao.org/search/en/?cx=018170620143701104933%3Aqq82jsfba7w&q=BAST (accessed May 1, 2016).
4. Dammer, L.; Carus, M.; Raschka, A.; Scholz, L. Market Developments of and Opportunities for bio based products and chemicals, Final Report. [Online] 2013, nova-Institute for Ecology and Innovation http://www.nova-institute.edu (accessed May 15, 2015).
5. Languri, E.; Moore, R. D.; Masoodi, R.; Pillai, M.; Sabo, M. An Approach to Model Resin Flow through Swelling Porous Media made of Natural Fibers. The 10th International Conference on Flow Processes in Composite Materials (FPCM10), Monte Verità, Ascona, CH. Switzerland, 2010.

6. Hoa, M.; Wanga, H.; Lee, J.; Ho, C.; Lau, K.; Leng, J.; Hui, D. Critical factors on manufacturing processes of natural fiber composites. Composites Part B 2012, 43(8), 3549–3562.
7. Cicala, G.; Cristaldi, G.; Recca, G.; Latteri. A. Composites Based on Natural Fiber Fabrics, Woven Fabric Engineering, Dubrovski, P. D. (Ed.). 2010.
8. Timber network, New study reveals the importance and development potential of the Biocomposite market. http://www.fordaq.com/fordaq/news/Biocomposites_wood_natural_fiber_composites_nova_35056.html (accessed May 7, 2016).
9. Oksana, K.; Skrifvarsb, M. Selinc, Natural fibers as reinforcement in Polylactic acid (PLA) composites. Compos. Sci. Technol. 2003, 63, 1317–1324.
10. Lightweight structures BV, Natural Fiber Composites – from upholstery to structural components, http://www.lightweight-structures.com/natural-fiber-composites-from-upholstery-to-structural-components/index.html (accessed May 8, 2016).
11. Chandramohan, D. Studies on natural fiber particle reinforced composite material for conservation of natural resources. Adv. Appl. Sci. Res. (Online) 2014, 5(2), 305–315. www.pelagiaresearchlibrary.com (accessed May 1, 2016).
12. Sanjay, M. R.; Arpitha, G. R.; Naik. L. L.; K. Gopalakrishna, K.; Yogesha, B. Applications of Natural Fibers and Its Composites: An Overview. Nat. Resourc. Sci. Res. [Online] 2016, 7, 108–114. http://www.scirp.org/journal/nr (accessed Apr. 12. 2016).
13. Quadrennial Technology Review 2015 Chapter 6: Innovating Clean Energy Technologies in Advanced Manufacturing Technology Assessments. University of America department of energy [Online] 2015 http://energy.gov/sites/prod/files/2015/12/f27/QTR2015-6E-Composite-Materials.pdf (accessed April. 12. 2016).
14. Zhu, J.; Zhu, H.; Njuguna, J.; Abhyankar, H. Recent Development of Flax Fibers and Their Reinforced Composites Based on Different Polymeric Matrices. Mater. 2013, 6, 5171–5198.
15. Cullen, R. K.; Margaret Singh, M. M.; Summerscales, J. Characterization of Natural Fiber Reinforcements and Composites. J.Compos. [Online] 2013, 1–4. http://www.hindawi.com/journals/jcomp/2013/416501/ (accessed May 1, 2016).
16. Summerscales, J.; Le Duigou, A.; Baley, C. A checklist for the description of plant/vegetal/bast natural fiber reinforced composites. Proceedings of 20th International Conference on Composite Materials, Copenhagen, Denmark, 2015.
17. Simmons C. T.; Henry Darcy (1803–1858): Immortalised by his scientific legacy. [Online] http://www.envsci.rutgers.edu/~gimenez/Soils%20and%20Water-Lectures/Darcy%20Scientific%20Achievement.pdf (accessed May 1, 2016).
18. Amico, S.; Lekakou, C. Mathematical modeling of capillary micro-flow through woven fabrics. Composites Part A 2000, 31, 1331–1344.
19. Rodríguez, E. S.; Francucci, G.; Vázquez, A. Study of saturated and unsaturated permeability in natural fiber fabrics. The 9th International Conference on Flow Processes in Composite Materials, Montréal (Québec), Canada, 2008.
20. Francucci, G.; Vázquez, A. Key differences on the compaction response of natural and glass fiber preforms in liquid composite molding. Text. Res. J. 2012, 82(17), 1774–1785.
21. Ferland, P.; Guittard, D.; Trochu, F. Concurrent Methods for Permeability Measurement in Resin Transfer Molding. Polym. Compos. 1996, 17(1), 194–158.

22. Oksman, K. High Quality Flax Fiber Composites Manufactured by Resin Transfer Molding Process. J. Reinf. Plast. Compos. 2001, 20(7), 621- 627.
23. Olivero, K. A. Effect of Preform Thickness and Volume Fraction on Injection Pressure and Mechanical Properties of Resin Transfer Molded Composites. J. Compos. Mater. 2004, 38(11), 937–957.
24. Kaynak, C. Effects of RTM Mold Temperature and Vacuum on the Mechanical Properties of Epoxy/Glass Fiber Composite Plates. J. Compos. Mater. 2008, 42(15), 1505–1521.
25. Diallo, M. L.; Gauvin, R.; Trochu, F. Key Factors Affecting the Permeability Measurement in Continuous Fiber Reinforcements. Proceedings of 23rd International Conference on Composite Materials, ICCM-11, Gold Cost, Australia, 1997.
26. Cherouat, A.; Borouchaki, H. Present State of the Art of Composite Fabric Forming: Geometrical and Mechanical Approaches. Materials. [Online] 2009, 2, 1835–1857 file:///C:/Users/magdy/Downloads/materials-02–01835%20(6).pdf (accessed May 6, 2016).
27. Umer, R.; Bickerton, S.; Fernyhough, A. Characterization of wood fiber mats as reinforcement for the resin transfer molding process. Proceedings of the 8th International Conference on Flow Processes in Composite Materials (FPCM-8), Douai, France, 2006, 97–104.
28. Vazquez, A.; Ruiz, E.; Francucc, G.; Rodrıguez, E. Key factors affecting processing of natural fiber composites, Plastic research. Soc. Plast. Eng. [Online] http://www.4spepro.org/pdf/004524/004524.pdf (accessed Aug. 15, 2015).
29. Francucci, G. M.; Rodriguez, E. S.; Vázquez, A. Study of the compaction behavior of natural fiber reinforcements in liquid composite molding processes. The 10th International Conference on Flow Processes in Composite Materials (FPCM10) Monte Verità, Ascona, CH, ETH Zurich, Switzerland, 2010.
30. Mastbergen, D. B. Simulation and testing of resin infusion manufacturing processes for large composite structures. MSc. Thesis, Montana state university-Bozeman, 2004.
31. Jaksic, D.; Jaksic, N. Assessment of Porosity of Flat Textile Fabrics. Text. Res. J. 2007, 77, 105–111.
32. Zupin, Z.; Hladnik, A.; Dimitrovski, K. Prediction of one-layer woven fabrics air permeability using porosity parameters. Text. Res. J. 2012, 82, 117–128.
33. Ravirala, N.; Gong, R. H. Effects of Mold Porosity on Fiber Distribution in a 3-D Nonwoven Process. Text. Res. J. 2003, 73, 588–592.
34. Robertson, A. F. Air Porosity of Open-Weave Fabrics Part I: Metallic Meshes. Text. Res. J. 1950, 20, 838–844.
35. Polona Dobnik Dubrovski, P. D.; Miran Brezočnik, M. The Usage of Genetic Methods for Prediction of Fabric Porosity, Genetic Programming – New Approaches and Successful Applications. Soto, S. V. (Ed.), InTech, [Online] 2012 http://www.intechopen.com/books/genetic-programming-new-approaches-and-successful-applications/ (accessed Dec.12, 2015).
36. Zaepernick, N. Industrial 3-D Printing – from Rapid Prototyping to Production. http://www.photonics21.org/download/Annual_Meeting/AnnualMeeting2013/Presentations/2013–04–30_From-Rapid-Prototyping-to-Production_presented.pdf (accessed Dec.12, 2015).

37. Amico, S.; Lekakou, C. An experimental study of the permeability and capillary pressure in resin-transfer molding. Compos. Sci. Technol. 2001, 61, 1945–1959.

38. Goebner, J. A Peek into the EOS Lab: Micro Laser Sintering, e-manufacturing solutions. https://scrivito-public-cdn.s3-eu-west-1.amazonaws.com/eos/public/edd2302b-d9ae390e/c51e3234355cfc4834e0ecca4f70ed9a/eos_microlasersintering.pdf (accessed Dec.8, 2015).

39. Vacuum Infusion-The Equipment and Process of Resin Infusion. [Online] http://www.composites.ugent.be/home_made_composites/documentation/FiberGlast_Vacuum_infusion_process.pdf (accessed Dec.8, 2015).

40. Udroiu, R.; Nedelcu, A. Optimization of Additive Manufacturing Processes Focused on 3-D Printing chapter 1. Rapid Prototyping Technology – Principles and Functional Requirements. Hoque, M. (Ed.). 2011 [Online] http://www.intechopen.com/books/rapid-prototyping-technology-principles-and-functional-requirements/optimization-of-additive-manufacturing-processes-focused-on-3d-printing (accessed Dec.8, 2015).

41. Daneshmand, S.; Aghanajafim, C. Description and Modeling of the Additive Manufacturing Technology for Aerodynamic Coefficients Measurement. Strojniški vestnik. J. Mech. Eng. 2012, 58(2), 125–133.

42. Daneshmand, S.; Nadooshan, A. A.; Aghanajafi, C. Investigation of Layer Thickness and Surface Roughness on Aerodynamic Coefficients of Wind Tunnel RP Models. Proceedings of World Acad. Sci. Eng. Technol. 2007, 26, 7–12.

43. University of Northernlowa, Major RP Technologies. [Online] http://www.uni.edu/~rao/rt/major_tech.htm#LENS (accessed Dec.8, 2015).

44. Printing with wood filament. http://www.tridimake.com/2012/10/review-wood-filament.html

45. 3-D printer. http://www.3ders.org/articles/20121105-finding-the-proper-temperature-ranges-for-3d-printing-wood.html

46. Advanced Manufacturing Choices. [Online] http://mmadou.eng.uci.edu/classes/mae165/.../ENG165–265.5.%202015.ppt (accessed Dec, 12, 2015).

47. Udroiu, R.; Ivan, N. V. Rapid Prototyping and Rapid Manufacturing Applications at Transilvania University of Braşov. Bulletin of the Transilvania University of Braşov. 2010, 3(52), Series I: Engineering Sciences.

48. El-Haggar, S. M.; Kamel, M. A. Wood Plastic Composites, Chapter 13, Advances in Composite Materials – Analysis of Natural and Man-Made Materials. Tesinova, P. (Ed.), INTECH 2011 [Online] http://www.intechopen.com/books/advances-in-composite-materials-analysis-of-natural-and-man-made-materials/wood-plastic-composites (accessed Dec 30, 2015).

49. 3-D printer and 3-D printing news. Could wood-based material lead 2014 3-D printing priorities http://www.3ders.org/articles/20131223-could-wood-based-material-lead-2014-3d-printing-priorities.html

50. Hietala, M. Extrusion Processing of Wood-Based Biocomposites. PhD. Thesis, University of Oulu, Finland, 2013.

51. Wang, J. Chapter 1, PVT Properties of Polymers for Injection Molding, Some Critical Issues for Injection Molding. [Online], Ed. Wang J., INTECH 2012. http://www.intechopen.com/ (accessed Dec. 20, 2015).

52. Lewandowski, K.; Zajchowski, S.; Mirowski, J.; Kościuszko, A. Studies of processing properties of PVC/wood composites. Chemk. 2011, 4, 333–336.
53. Gardner, D. J. Extrusion of Wood Plastic Composites. [Online] http://www.entwoodllc.com/PDF/Extrusion%20Paper%2010–11–02.pdf (accessed Dec. 20, 2015).
54. Farsi, M. Chapter 10, Thermoplastic Matrix Reinforced with Natural Fibers: A Study on Interfacial Behavior, Some Critical Issues for Injection Molding. Ed. Wang J. INTECH 2012. [Online] http://www.intechopen.com/ (accessed Dec. 20, 2015).
55. Material safety data sheet wood flour http://www.rotdoctor.com/products/msdspdf/Wood_Flour.pdf (accessed Dec. 25, 2015).
56. Potter, K. D. Understanding the origins of defects and variability in composites manufacture. Proceeding of ICCM-17, 2009, Edinburgh, UK. http://www.iccm-central.org/Proceedings/ICCM17proceedings/Themes/Plenaries/P1.5%20Potter.pdf (accessed Dec. 30, 2015).
57. Smith, R. Composite defects and their detection, Material since and engineering. [Online] 1996, 3, 103–144. http://www.eolss.net/sample-chapters/c05/e6-36-04-03.pdf (accessed May 5, 2016).
58. Mesogitis.T. S.; Skordos, A. A.; Long, A. C. Uncertainty in the manufacturing of fibrous thermosetting composites: A review. Composites Part A 2014, 57, 67–75.
59. Summerscales, J. Manufacturing defects in fiber-reinforced plastics composites. Insight, 1994, 36(12), 936–942.
60. Turoski, L. Effects of manufacturing defects on the strength of toughened carbon/epoxy prepreg composites. MSc Theses, Montana State University – Bozeman, Bozeman, Montana, USA, 2000.

CHAPTER 7

AGRICULTURE WASTE COMPOSITES

CONTENTS

7.1   INTRODUCTION

The earliest composite materials were straw reinforced bricks, which was similar to modern steel reinforced concrete [1]. The agriculture waste is one of the problems of the modern intensive agriculture which needs innovative methods for its useful industrial applications. The development of composites from renewable raw materials has increased extensively during the last years as they are considered to be environmentally friendly materials. Natural fibers are renewable, easily recycled, carbon dioxide neutral, and are available in large quantities. In addition, they have high specific properties and cause less health problems during handling when compared to glass fibers. Currently, natural fibers, mainly flax and hemp, are available as commercial products as short fibers for injection molding compounds or as non-woven mats based on short fibers organized in a random array. Enhancing the adhesion between the natural fibers and matrix results in improving composite strength and toughness [2–6]. Green composite combines plant fibers with biodegradable resins to create natural composite materials. Biomaterial composites are made from hemp, kenaf, sisal, soybean, and agro-residuals. Natural fibers are emerging as low cost, lightweight and apparently environmentally superior alternative to synthetic fibers [7]. With natural fiber composites, car weight reduction is possible. This can be translated into lower fuel consumption and the lower environmental impact. Natural fiber based composites also offer good mechanical performance, good formability, high sound absorption and cost savings due to low material costs. Moreover, their "Green look" as well as ecological and logistical benefits of the natural fiber based technologies looks more attractive. In 2000, more than 23,000 tons of natural fibers have been used in the automotive sector alone, furthermore, natural fibers in automotive industry should experience a sustainable growth EU [3].

There are currently massive unused quantities of agricultural (straw) residues around the globe. The estimated worldwide availability of wheat and rice straw in several countries is shown in Figure 7.1. The world available wheat straw reaches 583,776 tons and the rice straw 727,400 tons [8]. China, India and the USA appear to be at the present the major producing countries of straw residues (mainly wheat and rice straw). The harvest index of different crops is given in as: wheat and barley 0.55, rice 0.5, maize 0.52, sun flower 0.5, cotton 0.33, sugar beets 0.5. The wheat and rice straw are not the only agro residuals but also the other grain crops and cotton increase the amount of available materials for WPC. Figure 7.1 shows the availabilities of the other straw residuals.

According to the statistical data, the average yield of straw is around 1.3–1.4 kg per kg of grain [4]. The world statistics of the wheat and rice straw reveals that there are about 710 million metric tons of wheat straw and 650–975 million ton of rice straw each year [5]. The forest residuals also represents a source of material which can be used for composite production. The forest residuals may reach 20–35%, besides wood industry waste and the other unused residues (Figure 7.2).

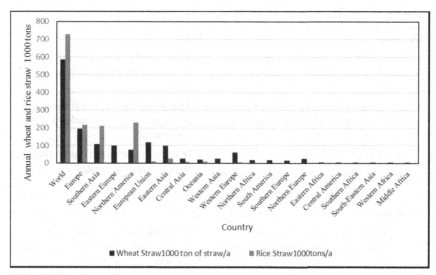

FIGURE 7.1 Comparison of the average annual wheat and rice straw in some countries [8].

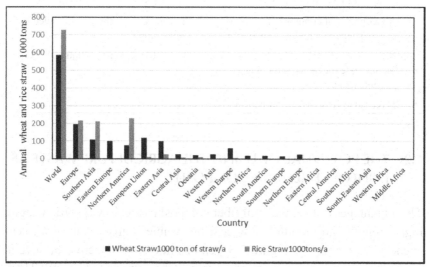

FIGURE 7.2 Agriculture waste [9].

The SWOT assessment of various aspects of rice straw and wheat straw, when used as WPC, discloses the far-reaching prospective of using straw of different agriculture residuals, as given in Table 7.1.

TABLE 7.1   SWOT Analysis of Wheat and Rice Straw Applications

Strengths	*Weaknesses*
• Rice straw and wheat straw are available in many countries around the world	• There are high costs associated to collection, handling, and transport of straw
• Rice straw and wheat straw are the most abundant agricultural residues in the world	• The required storage facilities and pre-treatment process makes straw less attractive
• Straw is a "Non-food" feedstock	• In many countries the supply chain of straw is very fragmented
• Straw exhibits a high cellulose content	
• Positive environmental impact	
Opportunities	*Threats*
• Increased grain production in the world leads to more straw being produced	• Other non-energy uses of straw compete with straw use for bio based economy
• Increased Environmental legislative efforts to stop open field burning of straw will make straw available	• Implementation of Sustainability criteria might lead to higher extraction rates
• Development of composite technologies may lead to a higher demand for straw	• The increase of the cost of collection and storage
• The increase the demand on biodegradable materials	• Limitations of application of WPC
• Increased environmental awareness	

7.2   THE POTENTIAL OF USING THE AGRICULTURE WASTE AS REINFORCEMENT OF COMPOSITE

The advantages of using natural fiber composites are: low specific weight, high strength and modulus, low cost renewable source of material, eco-friendly, easy recycle. Figure 7.3 demonstrates the objective of several researchers to turn the agro-residual into green composite [10–19]. There are huge unused quantities of agricultural (straw) residues around the globe. In 2014, the annual global production of wheat straw was around 705 million tons, out of which 75% was produced by 18 countries, while around 20% was produced by EU [7].

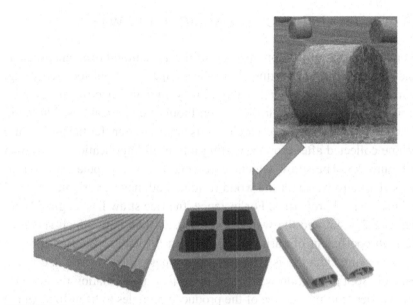

FIGURE 7.3   From rice straw bale to WPC profiles.

Increased environmental awareness and societal needs serve as a catalyst for developing new eco-friendly materials, like green composites. Agro-residuals and Biodegradable resin, starch based biodegradable polymer, soy flour resin – either can be used for the production of several items.

Thermoset composites or thermoplastic are commonly used in several applications, such as particleboard, fiberboard (MDF, HDF, cardboard, hardboard), wood-natural fiber composites, and non-woven mats. These products found applications in construction, automotive and marine industries, electrical components, appliances and furniture. WPC are used in decking and exterior and high density fiber (HDF) may be used in some flooring laminates. Wood-polymer composites (WPC) have gained a lot of concern in the last decade due to the optimization and development of machinery, material, and processing conditions. Polymer composites made with wood and natural fibers, such as rice hulls, flax, hemp, jute, kenaf, agricultural residue are environmentally sound due to their biodegradability [1, 7–9]. More than 350,000 tons of natural fibers composites are expected to be used in EU in 2015, mostly are WPC, over 80,000 tons used in automobile industry alone.

7.3   PRE-TREATMENT OF THE AGRICULTURE WASTE

There are several utilization methods of the agricultural residuals, such as landfill, industrial composting, domestic composting, feedstock recycling, animal bedding, biogas plant, thermal recycling and electric energy generation, as well as in bioplastic or non-bioplastic applications. The main direction of the scientific research in this area is to transfer the agriculture residue collected after crop harvesting into useful applications, as shown in Figure 7.3. The straw, which represents the most popular material to be used as agro waste, or the wood residue, both have a variable physical and mechanical properties. For instance, the rice straw has length 51–66, diameter 2.2–4.8 and the wall thickness 0.23–0.68 mm, which specifies the high coefficient of variations, making it difficult to be used directly [8]. The same situation is with the other types of agro waste and wood residual. So, the solution was found in milling or fibrillation these material in order to unify the size of the produced particles to be utilized in the composites. Also, the chemical analysis of the different types of agriculture waste point out that they have different percentage of cellulose, lignin, length and width. Table 7.2 demonstrates some lignocellulosic fibers.

TABLE 7.2   Chemical Composition and Properties of Agro Residuals [17–22]

Type of Fiber	Cellulose [%]	Lignin [%]
Bagasse	32–37	18–26
Cereal-straw	31–45	16–19
Corn-straw	32–35	16–27
Wheat-straw	33–39	16–23
Rice-straw	28–36	12–16
Cotton stack	45 -49	11- 27
Bamboo	26–43	21–31
Sugarcane bagasse	42	20
Nut shells	25–30	30–40
Grasses	25–40	10–30
Oil palm press	40	24
Wood-based Softwood	40–45	26–34
Hardwood	38–49	23–30

Generally, the length of straw fibers is shorter than softwood fibers, but is in the same level as hardwood fibers. The amount of lignin and cellulose is also lower than in both hardwood and softwood fibers. Hemicellulose (%) is varied between 20–40%. Wheat and rice straw, both are of the same composition, with about 39% of cellulose.

There are several preparation methods applied to the agro waste which depend on the nature of the waste itself and the required end product characteristic. Before applying the pre-treatment process, the following steps should be followed:

1. Selecting the classified fiber materials according to a desired property;
2. Choosing agro waste fiber/husk composition;
3. Selecting proper preparation technology for a specific agro-waste;
4. Defining the suitable additives, coupling agent and polymer;
5. Choosing the proper manufacturing technology.

All the above steps represent one package for the successful production of wood polymer composite.

7.3.1 PRE-TREATMENT TECHNOLOGIES OF STRAW AND WOOD RESIDUALS

The objective of the pre-treatment of the straw of agro residual is to transfer the low long density straw into high-density fibers of short lengths suitable to manufacturing into thermoplastic composite. The application of pre-treatments breaks the long-chain hydrogen bond in cellulose, making hemicelluloses amorphous, loosening the lignin out of the lignocellulosic matrix and reducing its percentage, resulting in better quality to make it more suitable for various applications. These methods can be classified as: physical, physicochemical, chemical, and biological [17, 23–25]. The result of the process of breaking down the stem into fiber element makes it possible to transport and handling of the agro residuals. The bulk density of rice straw is around 75 kg/m^3 for loose straws, 100–180 kg/m^3 in packed and in the bale form, and it increases in the pellet form 560–720 kg/m^3 [26]. Also, it allows mix easily with a polymer when using one of the forming techniques to manufacture the various products and facilitates easy and economical storage.

7.3.2   PHYSICAL METHODS (PHM)

Usually, the mechanical means are applied to reduce the stem into smaller size, such as milling, grinding, and chopping. Figure 7.4 shows a sketch

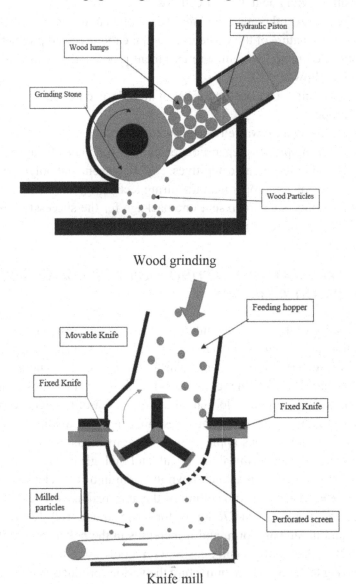

Wood grinding

Knife mill

FIGURE 7.4   Sketch of the principles of machine grinder and knife mill.

of milling and grinding machine principle. The size of the particle will be in the range of 0.2–2 mm milling, 10–20 mm chopping. More than one operation can be applied, starting with chopping then milling and sieving, to achieve the length as required (>5 mm to <0.1 mm in length). Milling is generally done by mechanical means, including abrasion, compression, and impact. Impact mills include the ball mill, which has media that tumble and fracture the material. Shaft impactor causes particle-to-particle attrition and compression [4]. Several passes are needed to reach the required size.

The material size can be classified according to the size of the particles through sieving with different mesh size as shown in Figure 7.5.

7.3.3 PHYSICO-CHEMICAL METHODS (PHCM)

Steam explosion; in this process high-pressure saturated steam is injected into a reactor filled with agro residuals. During the steam injection, the temperature rises to 160–260°C. Successively, pressure is suddenly reduced and the agro residuals undergo an explosive decompression with hemicellulose degradation and lignin matrix disruption as a result. During steam explosion pre-treatment process, the lignocellulose agro residuals are heated with high-pressure saturated steam having high temperatures for 1–15 minutes [27]. Subsequently, the material is quickly subjected to atmospheric pressure: as a result, the water inside the material vaporizes and expands rapidly, disintegrating the agro residuals. This causes great reduction in the particle size of the straw. Consequences of steam-explosion pre-treatment depend on treatment time, temperature, particle size and moisture content [28]. Investigations have been carried out to try to improve the results of steam explosion by addition of chemicals such as

| Hopped only | chopped and milled | sieved | milled wood flour |

FIGURE 7.5 Wood particle's classification.

acid or alkali [29]. The temperature of steam and the duration time may be reduced. The steam explosion process offers several attractive features when compared to other technologies [30]. These include less hazardous process chemicals and significantly lower environmental impact. Steam explosion process is a cost effective method for hard wood and agriculture residuals pre-treatment. There are several physico-mechanical methods, such as AFEX (Ammonia fiber explosion method), ARP (Ammonia recycle percolation method), CO_2 explosion, Ozonolysis and Wet oxidation. All required to increase the temperature (90°C to 200°C), except Ozonolysis, and time of treatment (30- 90 min), more than that of steam explosion process (1–15 min) [31].

7.3.4 CHEMICAL METHOD (CM)

The straw can be treated by Acid, either diluted or concentrated, Alkali (sodium hydroxide, calcium hydroxide, potassium hydroxide, Sulphur dioxide, ammonia, etc.) or Oxidization agent (H_2O_2). This treatment not only reduces the size of particles but also affects the pore size, lignin and hemicellulose. Depending on the type of chemical used, pre-treatment could have different effects on structural components. Alkaline pre-treatment, ozonolysis, peroxide and wet oxidation pre-treatments were reportedly more effective in lignin removal, whereas diluted acid pre-treatment was more efficient in hemicellulose solubilization [32, 33]. Ozonolysis pre-treatment is carried out at room temperature and normal pressure. It can effectively remove the lignin without producing any toxic residues or affect cellulose and with slight influence on the hemicellulose. The chemical treatment of straw may be:

   Acidic Pretreatment: diluted acids will treat the stacks at high temperature 160°C or strong acid hydrolysis, which need lower temperature. H_2SO_4 and HCl, are used. The straw will be treated by enzymes only in the case of diluted acid treatment.

   Alkaline Pretreatment: alkaline solutions, such as calcium, potassium, sodium and ammonium hydroxide are used for alkali pre-treatment of corn Stover, switchgrass, bagasse, wheat, rice straw, hardwood, and softwood. Pre-treatment doesn't require elevated temperature and gives a satisfactory results.

Wet Oxidation: dry milled straw is heated for 20min in temperature up to 195°C, followed by the addition of water and caustic soda. The application of air for oxidation leads to lignin removal. This method can be used with all agro-residuals [34].

Ionic Liquids: this method introduces lignin solvents to the agro-residual at temperature 90–130°C.

Oxidative Delignification: the agro-residual is treated with oxidizing agents, such as ozone, hydrogen peroxide. Ozonolysis oxidations will dissolve lignin only.

Organosolv: organic solvent such as ethanol, methanol, acetone are applied at temperature 200°C.

The most important and widely used methods for cellulose Nano fibers isolation are chemical methods, mechanical methods, physical methods, using microwave, biological and high-pressure homogenizer. CM process is easier than milling process (MM) by which material is reduced from a large size to a smaller size 'top-down.' A combination of the above methods is usually applied.

7.3.5 BIOLOGICAL METHODS (BOIM)

Using enzymes, may be one potential alternative to provide a more practical and environmental-friendly approach for enhancing the nutritive value of rice straw. The approaches are costly and also affect lignin and hemicellulose [35, 36]. Enhancement of the biological pre-treatment was suggested by several researchers for delignification of agricultural residues [37] through the chemical pre-treatment by alkali or diluted acid pre-treatment. White-rot fungi and Soft-rot fungi are capable of degrading lignin without affecting much of cellulose and hemicelluloses. Several types of fungi can be used with different straw type, such as Aspergillus Niger, Aspergillus awamori, Trichoderma reesei, Phanerochaete chrysosporium, Bjerkendra adusta, and Cyathus stercoreus. Combination treatment of the agro residuals by mechanical, chemical and biological methods is usually used to get the best result for the pre-treatment process of the different types of straw and wood residuals. For instance, in the pre-treatment process of wheat and rice straw, mechanical, physical and chemical methods are applied all together [38, 39]. Diverse pre-treatment methods were outlined in terms of the mechanisms involved, the choice of the optimal pre-treatment process depends greatly on the objective of the agro residual end-use. Furthermore, the selection of a pre-treatment method should not only be based on its final profit but also on the environmental impact. Figure 7.6 shows the effect of fungi treatment of wheat straw.

7.4   PREPARATION OF AGRICULTURAL RESIDUES

7.4.1   PRODUCTION OF WOOD FILLER

The WPC use wood that is derived from residues of the forest industry recycled wood or sawdust. Wood residuals require to be processed both mechanically and chemically. The mechanical processing purpose is cutting and milling the wood fiber to desirable size. For composite material preparation, it is necessary to remove from fibers fine organic impurities and greases and modify their length. Optimal length of milled fibers ranges within 0.5 up to 2 mm.

The chemical treatment reduces moisture absorption and enhances the wood-plastic bindings. Whatever is the treatment type of the wood, there is a need for wood to be dried to decrease the moisture content that reduces the wood – plastic interfacial bonding. The wood filler can be either in the shape of flakes, fibers, chips or flour and the size of the flour or fiber depends on the manufacturing technique and desired characteristics of the finished product. To decrease the moisture content in wood, one of the following treatments can be applied: acetylation, thermal modification and furfurylation.

Straw   After fungi treatment

FIGURE 7.6   Fungi treatment of straw.

- *Acetylation:* Acetylation is a process whereby one hydroxyl group is replaced by one acetyl group through impregnating the wood with acetic anhydride and then exposing the impregnated wood to a higher temperature in order for the start the reaction.
- *Thermal Modification:* Thermal modification is process where the wood is subjected to a temperature around 200°C for several hours in a low oxygen atmosphere.
- *Furfurylation:* Furfurylation is a process where the wood is impregnated under pressure with furfuric alcohol. Furfurylated wood also exhibits improved mechanical properties and improved and low tendency to absorb moisture.

7.4.2 WOOD FLOUR

Wood flour is defined as a finely-ground wood cellulose. It is used in manufacturing a wide variety of products ranging from exterior composite decking/railing to office furniture to caster wheels. The number of wood/plastic composite applications continues to grow significantly every day.

Wood flour is produced from selected dry wood waste by several types of grinders and sized by mechanical or air screening methods. The characteristics of very small wood particles are dependent on the method of preparation. De-bonding the wood cells, mechanical methods leave the chemical nature of the wood unaltered, but in making crushed wood pulp, the fibers are separated by fragmentation of the wood to smaller size particles [29]. Blending wood flour with a mixture of plastic resins has resulted in the creation of durable, long-lasting composites that are less expensive and lighter in weight. Screening the wood fiber to 35 to 235 mesh size results in a particle size between 0.513- 0.064 mm. The particles from mesh size between 40 and 60 gives good flow characteristics and ease of mixing into the plastic matrix.

7.4.3 WOOD PLASTIC PELLETS (GRANULES)

The processes, as shown in Figure 7.7, starts by blending the meshed wood waste particles, according to the required proportion. The meshed wood

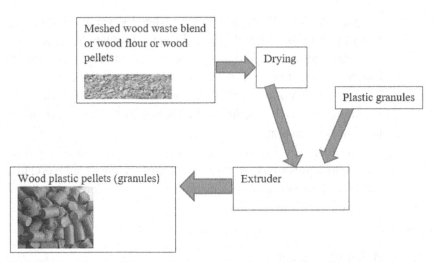

FIGURE 7.7   Sketch of the principle processing of wood plastic pellets (granules).

waste is dried in a furnace for 4 hours to almost completely eliminate the moisture. The furnace temperature is set at 115°C to avoid wood waste burning. Afterwards, dried wood waste is mixed with plastic and fed to the hopper of mixer and then fed into the single screw extruder. WPC additives are chemical components that facilitate wood-plastic composite processing (WPC extrusion) as well as/or impart WPC with improved properties. They come in form of powder, granules and, less often, liquids. Among the most common ones are coupling agents, compatibilizers, lubricants, UV-stabilizers, flame-retardants, antibacterial agents, antioxidants, etc.

Over forty coupling agents have been used in WPC [40]. Coupling agents are classified into organic, inorganic, and organic-inorganic groups. Organic agents include isocyanates, anhydrides, amides, imides, acrylates, chlorotriazines, epoxies, organic acids, monomers, polymers, and copolymers. Only a few inorganic coupling agents, such as silicates, are used in WPC. Organic-inorganic agents include silanes and titanates. A coupling agent accounts for only 1–3% of the total weight of a composite in WPC. Initiators are usually required with coupling agents during the coupling treatment, especially in graft copolymerization. The most widely used initiators are organic peroxides, including dicumyl peroxide, benzoyl peroxide, lauroyl peroxide, which accounts for only 0.5–1% of the total weight of a composite in WPC. The heating section of the extruder graduated between 120 to 150°C. In the extrusion line (One-step), the resulted extruded WPC are cooled and cut into the granules to predetermined specified dimensions.

7.4.4 WOOD PELLETS (GRANULES)

Wood pellets are usually made from sawdust, wood shavings, wood chips or wood logs, any forestry wastes or agro residuals straws, stalks, etc. Figure 7.8 shows the sequence of the wood pellet as:

Milling: the process starts with size reduction through milling of the different blend components;

Cleaning and Pre-Treatment: to remove the dust and any impurity and pre-treat the material by one of the discussed above techniques;

Drying: to achieve a material's consistent moisture level. Large dryer drums may use natural gas, propane, sawdust burners, or other fuels to heat the drum;

Pelletizing: after drying, the sawdust is pressed through dies at high pressure. This process causes the sawdust to heat up and release natural

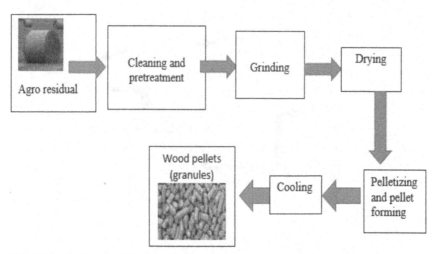

FIGURE 7.8   Sketch of the principle manufacturing of wood pellets (granules).

lignin in the wood that binds the sawdust together or to use extruders to give a small size pellets;

Cooling: cooling tower is used to bring the temperature down and harden the pellets.

The wood pellets can be blended with polymer and extruded to the required shape in the next step.

Estimated world wood pellet production is about 16 million tons, produced in EU, Canada, USA, almost 60% are produced in Canada only [28] using sawmill residues – sawdust, planer shavings, and other forest residuals.

7.4.5 WOOD RESIDUAL DEFIBRILLATION

The steps of the defibrillation are shown Figure 7.9. The wood residual defibrillation process involves size-reduction, screening, and washing of the raw material to get a clean and suitable size of the chips for consistent material flow into the Defibrator system [29]. The wood chips are pre-heated by steam and forced into a vertical preheater, then to the Defibrator system which is pressurized under pressure of 0.7–1.0 MPa

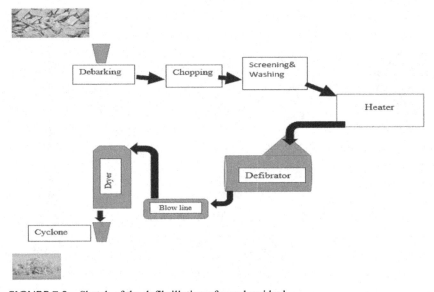

FIGURE 7.9 Sketch of the defibrillation of wood residual.

with temperature 170–190°C. The fibril material will pass through drying section ended by cyclone for further drying. The material is ready to be blended with a polymer and chemical additives and fed to the extruder.

7.4.6 WHEAT AND RICE STRAW

The wheat and rice straw after pre-treatment can be used for the formation of WPC in several forms as classified in Figure 7.10. The composite may be completely biodegradable when using matrix of bio gradable polymer [41–43].

Depending on the end use, there are several manufacturing technologies to deal with the rice and wheat straw: mechanical milling, chemical, biological method or combination of them. For the chemical pre-treatment of straw, different chemicals may be employed including acids, alkalis and oxidizing agents, such as peroxide and ozone. Among acid pre-treatment methods, diluted acid pre-treatment using H_2SO_4 is the most-widely practiced method [42]. For rice straw the best pre-treatment method is acid-catalyzed hydrolysis with high temperature and hot water [43]. Calcium, sodium, potassium and ammonium hydroxides are suitable alkaline agents for pre-treatment.

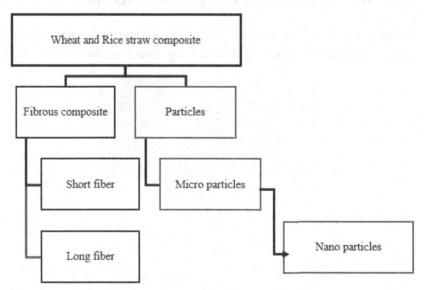

FIGURE 7.10 Classifications of wheat and rice straw composites.

For biological pre-treatment of straw, the most effective microorganism is white rot fungi, usually combined with other pre-treatment methods.

The other method is by using steam explosion. Pre-treatment of wheat straw starts by cutting it into chops of length in range of 10–50 mm. The size-reduced and dry straw material is later heated with steam and hot water in a screw mixer or soaked in water [17]. It is also possible to add chemicals at this stage. Afterwards, it is rapidly heated under pressure of 0.7–1.0 MPa to temperature 170–190°C for a period of time ranging in 1–3 min, then the pressure is suddenly reduced forcing the material undergo an explosive decompression. Figure 7.11 shows the change of the wheat straw after the steam explosion process.

Rice straw is fibrillated at a slighter lower pressure, approximately 0.6 MPa for 1 min. Rice straw will be soaked in a hot water before feeding to the Defibrator system [17]. Pre-treatment is an important step in biomass to conversion process and helps in overcoming the resistance of biomass for processing of NFPC [44].

7.5   NANO-SCALE CELLULOSE FIBERS

7.5.1   THE PRODUCTION OF NANO/MICRO FIBER

The production of Nano-scale cellulose fibers and their application in composite materials has gained increasing attention due to their high

Pretreated - cut wheat straw   Fibrillated wheat straw

FIGURE 7.11   Wheat straw fibrils after steam explosion.

strength and stiffness combined with low weight, biodegradability and renewability [45–51]. The nanotechnology has drawn much attention since the beginning of this century as the critical technology to advance industrial outputs. Developing commercial applications of nanotechnology, the Nano-sized particles play a significant role because of their unique functional properties. The nanoparticle technology is expected to further develop rapidly in all the industries handling fine particles in the near future [52–56]. Cellulose nanofibers have many advantages comparing with the inorganic fibers, some of the most relevant being the following: low density, renewable sources, biodegradability.

The main reason in using cellulose nanofibers in composite materials is attributable to the potentially high stiffness of the cellulose fibers for reinforcement. This property can be achieved with the reducing amount of amorphous material by breaking the structure of the plant into individualized nanofibers with high crystallinity [56].

Growing interest has been focused on biodegradable polymers based materials cellulose microfiber. One reason is the potential for significant improvements in the mechanical properties of fiber–plastic composites (FPC) for use as new materials due to the abundance, low cost, and its process ability. One of the famous applications is the use of wood as filler, wood flour, with a low length to-diameter (l/d) ratio 3.3 to 4.5. It was found that fiber length and distribution play important role in the processing and mechanical performance of fiber-based composites [57].

The mechanical properties of wood-plastic composites made with different percentages and sizes of pine flour in a HDPE matrix using an injection-molding process have been studied [58]. Compared to fiber proportion, size of screened particles showed no significant impact on mechanical properties of the composite materials. In terms of processing, it was suggested that mechanical properties and fiber length are sensitive to extrusion parameters, such as screw configuration and compounding temperature [59]. Short fibers resulted in better process ability than long fibers, water absorption and volumetric swelling increased with fiber length. Increasing fiber length had beneficial effects on tensile, flexural modulus of elasticity and toughness of wood-plastic composites. A decrease in the modulus of rupture was observed with the addition of wood fibers into the thermoplastic, but this effect could be minimized by using long fibers. The length of reinforced fibers plays a significant impact on some mechanical properties, modulus of elasticity, breaking elongation and the impact strength of fiber reinforced composites. Several studies emphasize the effect of fiber length and fiber orientation [60–62] on mechanical properties of the fabricated composites. The development of starch composite materials filled with Nano/micro sized rigid particles has attracted both scientific and industrial interest, in case of the rice starch (RS) films reinforced with microcrystalline cellulose from palm pressed fiber (MCPF). The incorporation of MCPF into rice starch films provided an improvement of water resistance for the rice starch films. In bio-composite films, rice starch film/MCPF increased tensile strength from 5.16 MPa, for pure rice starch film, to 44.23 MPa, but decreased in elongation at the break of composites [63]. Because of strong Nano fibril-polyurethane interaction at the interface, nanocomposites have large strain-to-failure. At high strains, cellulose Nano fibrils, as well as polymer molecules, become strongly reoriented in the loading direction, providing additional stiffening [64]. The production of MFC by fibrillation of cellulose fibers into Nano-scale elements requires intensive mechanical treatment. However, depending upon the raw material and the degree of processing, chemical treatments may be applied prior to mechanical fibrillation. These chemical processes are aimed to produce purified cellulose, such as bleached cellulose pulp, which can then be further processed. There are also examples with reduced energy demand in which the isolation of cellulose micro-fibrils involves enzymatic pre-treatment followed by mechanical treatments [46].

Micro fibrillated celluloses (MFCs) are generally obtained when cellulose fibers are submitted to high mechanical shearing forces. Nano fibrillated celluloses (NFPCs or Nano fibrils) are produced when specific techniques facilitating fibrillation are incorporated in the mechanical refining of wood and plants pulps. They ideally consist of individual nanoparticles with a lateral dimension around 5 nm [46].

Mechanical processes, such as high-pressure homogenizers, grinders/refiners crycrushing, high intensity ultrasonic treatments, and microfluidization have been used to extract cellulose fibrils from WF, PF, microcrystalline cellulose (MCC), tunicate, algae, and bacterial source materials. In general, these processes produce high shear that causes transverse cleavage along the longitudinal axis of the cellulose micro-fibrillar structure, resulting in the extraction of long cellulose fibrils, termed MFC. Typically, cellulose materials are run through the mechanical treatment several times (i.e., number of passes). After each pass, the particles are generally smaller, more uniform in diameter, but have increased mechanical damage to the crystalline cellulose. A filtration step is included to remove the larger unfibrillated and partially fibrillated fractions. In addition, these mechanical processes can be followed by chemical treatments to either remove amorphous material or chemically functionalize the particle surface [50]. Cellulose Nano-fibers would be prepared from MCC by application of a high pressure homogenizer (137.89 MPa) and treatment consisting of different passes (0, 1, 2, 5, 10, 15 and 20) [55].

The potential of cellulose Nano-fibrils as reinforcement provides a new direction for the development of value-added novel composites. These cellulose Nano-fibrils have been extracted from cell walls by a chemical [45, 50, 51] or mechanical treatment [65, 66], and combination of these treatments [69]. Cellulose whiskers from cellulosic materials were prepared by hydrochloric acid and sulfuric acid hydrolysis, and web-like cellulose Nano-fibrils were obtained by refining and homogenizing action. Cellulose Nano-fibrils with diameters below 100 nm were isolated by a chemical or mechanical treatment. Mechanical treatment with acid hydrolysis results in even finer cellulose Nano-fibril structures with diameters below 50 nm. MCC was used as raw material for the preparation of nanofibers, starting with MCC of about 20 um.

The mechanical treatment can produce particles in Nano/micro scale between 5 nm and 120nm from the different stacks, depending on the number of crushing passes and passes through the homogenization [67]. A fundamental mechanism of the high-pressure homogenizer is to bombard a fluid stream against itself within interaction chambers of fixed geometry at very high energy, resulting directly in the breakup and dispersion of the slurry. High pressure, high velocity and a variety of forces on the fluid stream are capable of generating shear rates within the product stream, reducing particles to Nano-scale.

The steam explosion process is another efficient pre-treatment method for converting lignocellulosic agro residuals with the final aim of separating nanofibers. In this process the agro residual is first milled and then subjected to high pressure steam for a short time (20 sec to 20 min) at a temperature 200–270∘C and a pressure of 14–16 bar. The pressure in the digester is then dropped quickly by opening the steam and the material is exposed to normal atmospheric pressure to cause explosion which breaks down lignocellulosic structure. Steam explosion causes the hemicelluloses and lignin from the wood to be decomposed and converted into low molecular weight fraction which can be recovered by extraction [68]. In order to fibrillate MCC, pass numbers were 1 up to 20, depending on the size required. The MCC fibrils have approximately the particle sizes of 10 um length and 2 mm diameter (aspect ratio of around 5). After further treatment to 20 passes, these small bundles were additionally split into thinner fiber bundles [55]. The effectiveness of the steam explosion is dependent on the agro residuals feedstock and, for instance, the process is less effective for softwood than for hardwood [57].

Other physical methods using gamma radiation or Altera sound wave are another approaches for straw pre-treatment. Microwave and radio frequency (RF) heating can be used instead of steam. The temperature of operation ranged from 70 to 230˚C, while heating time varied from 5 to 120 minutes, depending on the microwave power [69]. Alkali pre-treatment will help to reduce the time of fibrillation.

7.5.2   NATURAL NANO/MICRO FIBER FROM INDUSTRIAL WASTE

Cellulose particles can be obtained from the natural fiber waste extracted during the processing. The shaving method of staple fiber was suggested, resulting in particles of different diameters [70], Figure 7.12 illustrates the histogram of a sample of 700 particles. The value of cutting length is 2 micron. The analysis of the histogram of the sample particles length in micron evidences that 48% of the particles have a length less than 2 micron, which is the sum of residual length of the fibers with a tendency be skewed to the right. The mean value of particle length is 2437 nm and CV 2.87% with probability of Nano-particles presence, with law aspect ratio of 1.6 in average.

The analysis of the histogram of particle size equal to or less than 500 nm indicates that the average particle size is equal to 297 nm and CV 5%, as given in Figure 7.13.

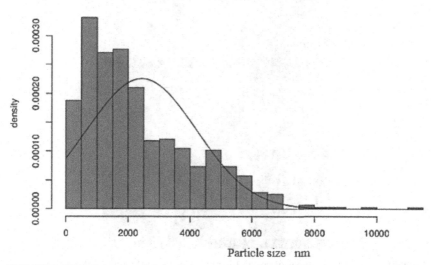

FIGURE 7.12   Histogram of the cellulose particles distribution.

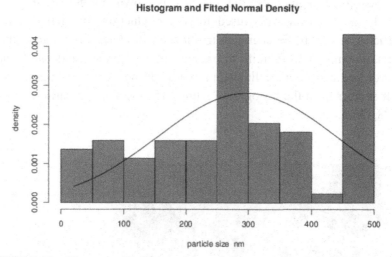

FIGURE 7.13   Histogram distribution of the cellulose particles of size equal to or less than 500 nm.

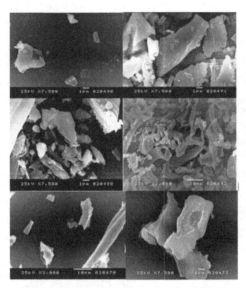

FIGURE 7.14   SEM images of cotton fiber particles using shaving methods.

The particle shapes produced by shaving method exhibit various dimensions of fragments of the particles, as shown in Figure 7.14.

The acid hydrolysis is used for the production of micro cellulose particles in which the chemical treatment will attack fiber's amorphous parts obtaining a more uniform particle length. Cotton fibers can be acid hydrolyzed and micro cellulose particles can be extracted Figure 7.15a, b. It is clear from the histogram Figure 7.15c the average particles size is 8208.33nm.

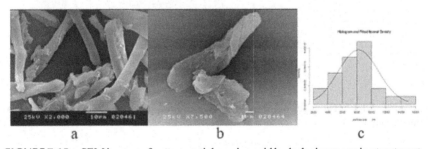

a   b   c

FIGURE 7.15   SEM images of cotton particles using acid hydrolysis processing treatment.

The length of the different fibers of cotton samples as well as the dimensions of the particles in each case denoted Table 7.3, categorizing that the smallest particle size was obtained when flat stripes or cotton wool were used, hence more than 50% of the particles had size less than one micron. Sisal plant and other fibers can be dried and ground into powder using a grinding machine and sieved to 150 μm and 300 μm sizes [69]. The sieved sisal powder was used as the reinforcement. The sisal powder and the ground polypropylene were blended for manufacturing of composite. Sisal particles improved the hardness property of the polypropylene matrix composite. The application of micro cellulose as a filling material change the characteristic of the PVA composite properties. Table 7.4 gives an example of using micro cellulose 20 μm in PVA matrix which indicates the increase of both Young's modulus and the shear modulus as the percentage of the micro particles increased while the failure strain decreases.

TABLE 7.3 Particle Length Frequency Distribution for Different Samples

Material type	Fiber length	Average particle size (μ) micron	CV %	Percentage of particle length < (μ) micron
Cotton	32	2.64	2.87	38%
Cotton (flats stripes waste)	10	1.99	8.9	50%
Cotton (taker-in waste)	18	2.53	6.43	32%
Cotton wool	12.5	2.41	8.92	40%
Cotton (ginning waste)	11	2.27	7.546	35%
Flax	25	2.53	6.436	32%

TABLE 7.4 Mechanical Properties of the MC/PVA Composite

MC/PVA Ratio after drying	Tensile stress N/cm^2	Strain%	Young's modulus E MPa	Shear modulus G MPa
0%	320.68	3.12	1.0287	0.459
6.33%	311.22	0.248	12.55	4.822
11.60%	353.22	0.135	26.16	9.69
16.20%	351.37	0.12	29.28	10.6
20.80%	383.29	0.15	25.55	9.07
24.80%	662.74	0.058	114.26	39.88

7.6   METHODS OF MANUFACTURING AGRO-WASTE COMPOSITES

After the pre-treatment of the agro-waste, the cellulosic material may be in the form of chopped material cut straw, fibers, particles, flour, nano/micro particles, according to the method used for pre-treatment and nature of the agro-waste. The agro-waste can be used in some products in raw form or after milling and mixed with the suitable polymer, PE, PP, PVC, adding other additives to manufacture pallets, packing plates, windows and door frames, decoration panels, etc. as illustrated in Figure 7.16.

There are several methods for manufacturing the pretreated agro residuals that cope with the final product shape and the nature of the filling materials used and finally, the costs.

7.6.1   EXTRUSION PROCESS

This process is common for thermoplastic material, with the advantages of high production rates, large volume of constitutes are produce. Raw materials, such as thermoplastic granules, pellets or powder, are placed into a hopper and feed into screw extruder [71]. Wood husk and straw can be processed with polyethylene (PP), polypropylene (PE) and polyvinylchloride (PVC) for the production of WPC using extrusion technique. Wood flour is mainly produced from the residues of various wood processes, and subjected to milling processes for size reduction. Wood plastic

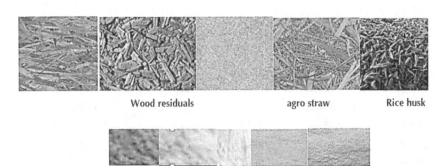

Wood residuals   agro straw   Rice husk

Wood flour   Rice husk powder

FIGURE 7.16   Agro material for manufacturing.

composites (WPC), made out of wood waste and plastic waste, have been a fast growing research area because of their wide range of application [72–75]. The manufacturing of wood-plastic composites starts with wood material preparation, blending with polymer, and molding. Various techniques were suggested for molding, injection molding, profile extrusion or compression molding. There are also wood pellets which provide pre-processed wood fibers as fillers for many applications. The fiber volume fraction may reach 30–40%.

However, the screw has four distinct sections: feed section, melt section, pressing section and forming section.

Molten polymer goes through a die to produce a final shape. It involves the following steps:

- Wood flour with different particle size, 74–500 μm [76], is blended in the required ratios;
- Wood flour is dried for 3–4 h at 105–115°C;
- Wood pellets and the polymer are mixed with coloring and additives (coupling agents, light stabilizers, pigments, lubricants);
- The granular plastic is blended with dry wood flour and fed in a hopper of extruder, as illustrated in Figure 7.17;
- Extruder screw thread turns forcing the wood-plastic material through a heater, melting it;
- The extrusion process carried out in the temperature ranging 140°C to 185°C that corresponds to the temperature of plastic granulates preparation and increases in the direction of the material flow;
- The wood-plastic material is forced through a die to produce the required shape;
- Cooling process is continued for a predetermine time.

Several types of polymers can be used in the extrusion processes, such as Polyethylene Terephthalate (PET or PETE), high density Polyethylene (HDPE), Polyvinyl Chloride or Vinyl (PVC-V), low density Polyethylene (LDPE), PP, Polystyrene (PS) and all have a potential for recycling. In the extruded WPC product, 50% use PP, 39 % use PE, and 15% use PVC [77]. The physical and mechanical properties of WPC depend on the wood fiber percentage content and type of a polymer. WPC with medium percentage of wood filler in PP or PE polymer have higher impact resistance, but lower strength and stiffness, while PVC polymer composites have high

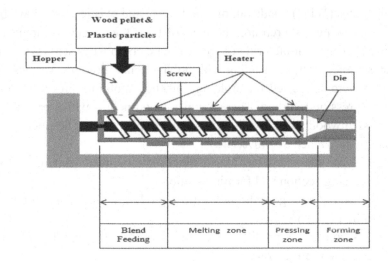

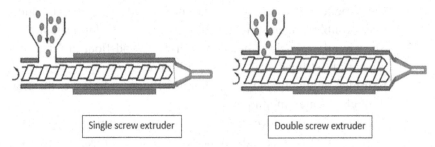

FIGURE 7.17   Principles of single screw and double screw extruder.

strength and stiffness and very low impact resistance with high thermal expansion.

Single screw extruder or double screw extruder can be used it is steadier, less heat required, exerting high pressure on the material and more in productivity.

7.6.2 INJECTION MOLDING

The sketch of the principles of the injection molding is given in Figure 7.18. The machine consist of two zones, the injector is where the first stage

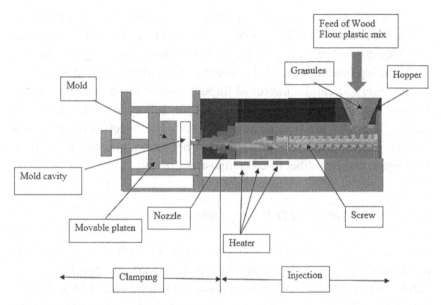

FIGURE 7.18 Injection molding machine.

of the injection molding process takes place. The injector is the part of the machine into which plastic or in the case of WPC pelletized compound is fed, molten, and then injected into the mold. As the screw rotates, it transports the molten WPC from the hopper towards the nozzle, it is transported further towards mold under pressure. The screw has three sections, for feed, melting and metering. The second zone is the clamping device. When the molten plastic is injected into the mold the mold needs to be held together in order for the pressure to be maintained. This is done by the clamping device [78]. The clamping device opens in the end of the cycle and ejects the finished product. The injection cycle consists of following steps, molting WPC injection, packing stage, ejection. The pressure and the temperature are varied during the injection cycle. The temperature is high 200°C at the start the dropped in the cooling period (most time of the injection cycle) till the injection is completed, while the pressure will be high at the injection and constant for the half time of the cycle, dwelling pressure after molten plastic fills out cavities, then dropped gradually till the injection of the molded part.

Polymers, such as polystyrene, nylon, polypropylene and polythene, can be used in the injection molding. WPC can be processed on the injection molding using polyethylene, polypropylene, and polystyrene. For WPC, strong binding is needed between hydrophilic wood and hydrophobic polymers. It can be improved through one or more coupling agents, Silanes, Carboxylated Waxes, Maleic Anhydride Polypropylene, DuPont's Fusabond, and/or pre-treatment of the wood. The injection molding process has been applied to wood fiber, flax fiber, hemp, jute, rice hulls, palm fiber waste and sisal fiber filled composites [78–80].

7.6.3 INJECTION BLOW MOLDING

Blow molding is a manufacturing process by which hollow plastic parts are formed, such as plastic bottles. This process forms the products from a variety of thermoplastic materials, including the following: LDPE, HDPE, PET, PP, and PVC. Depending on the materials, the blow ratio may be as high as 7:1. In extrusion blow molding (EBM), plastic is melted and extruded into a hollow tube. This tube is then captured by closing it into a cooled metal mold. Air is then blown into the tube (parison), inflating it into the shape of the hollow bottle, container or any part. After the plastic has cooled sufficiently, the mold is opened and the part is ejected [81]. Figure 7.19 illustrates the principles of Injection Blow Molding process. In blow molding, the composite formulation and the choice of the suitable polymer play a determined role in the quality of the product. For instance, softwood flour in the particle size range of 60 mesh, maple (hardwood) flour (80 and 140 mesh) was mixed with different grades of polypropylene. These grades included injection grade with moderate melt flow, blow molding grade with low melt flow and another blow molding grade with high melt strength. The matrix was filed at 25% wood flour [82]. In blowing molding the higher processing temperatures is required for pure polymers. Wood is understood to degrade at temperatures approaching 200–232°C, whereas plastics tend to perform optimally around this temperature range. Melting temperature of some polymers are PS 105°C, LDPE 105°C, HDPE 135°C, PP 165°C, and PVC 215°C [83]. Temperature effects the blow mold ability of a wood fiber reinforced plastic composite. Blowing is usually done with a hot air blast at pressure ranging from 350–700 kPa [71].

7.6.4 COMPRESSION MOLDING

Compression molding involves a pre-shape viscous mixture of liquid-resin and filler measures is place directly into a heated mold cavity at approximately 200°C. Forming is done under pressure from a plunge. The process is usually used in thermosetting plastics with the wood particles

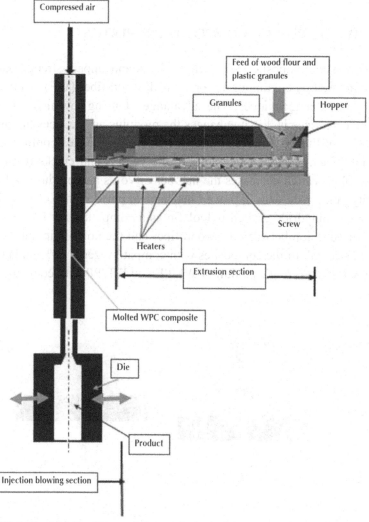

FIGURE 7.19 Injection blow molding method principles.

material being in a partially polymerized state. Figure 7.20 shows a sketch of the principle of compression molding.

The process of the manufacturing of the WPC depends on several parameters, such as the nature of the wood material, size of the particle, pre-treatment method, polymer properties, type of additives, shrinkage ratio, mold shape, applied pressure and temperature during processing and many other factors.

7.7 WOOD PLASTIC COMPOSITE PROPERTIES

Wood plastic composite with high modulus and impact strength can be manufactured by combining PE or PP with wood fiber (WF), using twin-screw extruder techniques. The advantage of using low melt viscosity polymer matrices is that it enhances the modulus and reduces the overall viscosity of the system. SEM analysis of the composites confirmed that the polymer molecules penetrate into the vessels and cracks of the cellulose fiber, which decreases the number of voids and produces a higher density composite with improved mechanical performance. The addition of maleic anhydride-grafted polyolefin as a compatibilizer improves the level of adhesion between the wood fiber and the polyolefin matrix. The impact strength of the composites with compatibilizer is 60% higher than those without. Young's moduli of WF/PE and WF/PP with compatibilizer

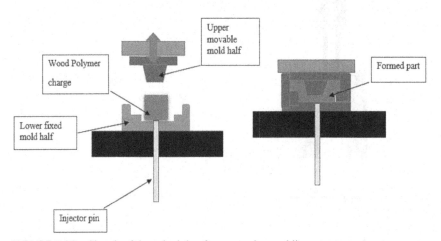

FIGURE 7.20   Sketch of the principle of compression molding.

were 4.4 and 5.4GPa, respectively, meanwhile the impact strengths of WF/ PE and WF/PP were 44 and 24J/m, where the WF content was 50 % [84].

The wood-plastic composite WPC are characterized as lightweight materials and easy to mold, allowing complex shapes. WPC end use covers the extensive variety products, especially to substitute the wood. The PP and polyester (PE) are widely used in manufacturing of WPC and the recycled polymers PP and PE [85]. WPC improves resistance to moisture and insects, and fungi attack. It was found that the composites with 30% wood sawdust also had higher water absorption, but water uptake is lower in composites with coupling agent, so this may have a protective effect against penetration of water. Innovative environmentally friendly processes for turning agricultural residues, like straw, into quality boards suitable for use in the furniture, construction, and the automotive and transportation industries are targeted by several researchers [86–88]. In general, ratio of 30% wood flour showed good physical and mechanical properties [88]. The polymers used with agro residuals are PE, PP, PVC, and PS with the coupling agents like Maleic anhydride and Hexamoll for PVC. The single or twin screw extruders are used, depending on the type of polymer [89, 90].

KEYWORDS

- agriculture waste
- biological methods
- chemical methods
- physical methods
- physico-chemical
- pre-treatment methods
- wood plastic

REFERENCES

1. El Messiry, M.; El Deeb, R. Study Wheat Straw Composite Properties Reinforced by Animal Glues as Eco-Composite Materials. Proceedings 2nd Conference for Industrial Textile Researches, National Research Centre (NRC), Cairo, Egypt, 2013.

2. Botros, M. Development of new generation coupling agents for wood-plasti composites. [Online] http://www.lyondellbasell.com/techlit/techlit/Tech%20Topics/Equistar%20Industry%20Papers/CouplingAgents%20for%20Wood-Plastic.pdf (accessed May 5, 2016).

3. Shanmugasundaram O. L. Green, Manufacturing Techniques and Applications. Indian J. T. 2009.

4. Goutianos, S.; Peijs, T.; Nystrom, B.; Skrifvars, M. Textile Reinforcements Based on Aligned Flax Fibers for Structural Composites. [Online] http://citeseerx.ist.psu.edu/viewdoc/download?doi=10.1.1.470.1201&rep=rep1&type=pdf, (accessed Dec. 12, 2015).

5. Pan, X.; Sano, Y. Fractionation of wheat straw by atmospheric acetic acid process. Bioresource Technol. 2005, Vol. 96, No. 11, 1256–1263.

6. Zhu, J.; Zhu, H.; Njuguna, J.; Abhyankar, H. Recent Development of Flax Fibers and Their Reinforced Composites Based on Different Polymeric Matrices. Materials [Online] 2013, 6, 5171–5198 http://www.mdpi.com/journal/materials (accessed Dec. 12, 2015).

7. Satyanarayana, K. G. Recent Developments in Green Composites based on Plant Preparation, Structure Property Studies. J. Bioprocess Biotec. [Online] 2015, 5:1–12 http://www.omicsonline.org/open-access/recent-developments-in-green-composites-based-on-plant-fibers-preparation-structure-property-studies-2155-(accessed Dec. 12, 2015).

8. NL Agency Ministry of Economic Affairs. Rice straw and Wheat straw. Potential feedstocks for the Biobased Economy [Online] 2013. http://english.rvo.nl/sites/default/files/2013/12/Straw%20report%20AgNL%20June%202013.pdf (accessed Dec. 12, 2015).

9. Kozlowski, R. Green Fibers and Their Potential in Diversified Applications. FOA, Trade and Markets Division [Online] 2000 http://www.fao.org/3/a-y1873e/y1873e0b.htm#fn31 38 (accessed Dec. 15, 2015).

10. Chevillard A.; Angellier-Coussy H.; Cuq B., Guillard V.; César G.; Gontard N.; Gastaldi E.; How the biodegradability of wheat gluten-based agromaterial can be modulated by adding Nano clays. Polym. Degrad. Stab. 2011, 96(12), 2088–2097.

11. Mads A.; Kristensen J.; Felby C.; Jorgensen H. Pre-treatment and enzymatic hydrolysis of wheat straw (Triticum aestivum L.) – The impact of lignin relocation and plant tissues on enzymatic accessibility, Technology. 2011, 102(3), 2804–2811.

12. Avella M.; La Rota G.; Martuscelli E.; Raimo M.; Sadocco P.; Elegir G.; Riva R. Poly(3-hydroxybutyrate-co-3-hydroxyvalerate) and wheat straw fiber composites: thermal, mechanical properties and biodegradation behavior. J. Mater. Sci. 2000, 35, 829–836.

13. Mantanis G., Nakos P., Berns J., Rigal L. Turning Agricultural Straw Residues into Value –Added Composite Products: A New Environmentally Friendly Technology. Proceedings of the fifth international conference on environmental pollution. Anagostopoulos, A. (Ed.), 2000, 840–849.

14. Yu, H.; Liu, R.; Shen, D.; Wu, Z.; Huang, Y. Arrangement of cellulose microfibrils in the wheat straw cell wall. Carbohydr. Polym. 2008, 72(1), 122–127.

15. Pearson C. Animal Glues and Adhesives, Handbook of Adhesive Technology, Ed. Pizzi, A.; Mittal, K. L., Marcel Dekker, Inc., 2003.

16. Fatoni R. Product Design of Wheat Straw Polypropylene Composite. PhD. Thesis, University of Waterloo, Canada, 2012.

17. Halvarsson, S. Manufacture of straw MDF and fiberboards. PhD Theses Mid Sweden University, Swede, 1992.

18. Kasmani, J. E.; Samarih, A. Some Chemical and Morphological Properties of Wheat Straw. Middle-East J. Sci. Res. [Online] 2011, 8(4), 823–825 http://www.idosi.org/mejsr/mejsr8(4)11/21.pdf (accessed March 5, 2016).

19. Abba, H. Nur. I.; Salit, S. Review of Agro Waste Plastic Composites Production, J. Miner. Mater. Charact. Eng. [Online] 2013, 1(5), 271–279 http://www.scirp.org/journal/jmmce (accessed May 1, 2016).

20. Bledzki, A. K.; Sperber, V. E.; Faruk, O. Natural and Wood Fiber Reinforcement in Polymers. A Rapra Review Report. Rapra Technol. Limit. 2002, 13(8).

21. Shaikh, A. J.; Gurjar, R. M.; Patil, P. G.; Paralikar, K. M.; Varadarajan, P. V.; Balasubramanya, R. H. Particle boards from cotton stalk. CIRCOT [Online] https://www.icac.org/tis/regional_networks/asian_network/meeting_5/documents/papers/PapShaikhA.pdf (accessed March 2, 2016).

22. Iqbal, H. M. N.; Ahmed, I.; Zia, M. A.; Irfan, M. Methods for pre-treatment of lignocellulosic agro residuals for efficient hydrolysis and biofuel production. Ind. Eng. Chem. Res. 2009, 48, 3713–3729.

23. Talebnia, F. Karakashev, K.; Angelidaki, I. Production of bioethanol from wheat straw: An overview on pre-treatment, hydrolysis and fermentation. Bioresource Technol. 2010, 101(13), 744–4753.

24. Bjerre, A. B.; Olesen, A. B.; Fernqvist, T.; Plöger, A.; Schmidt, A. S. Pre-treatment of wheat straw using combined wet oxidation and alkaline hydrolysis resulting in convertible cellulose and hemicellulose. Biotechnol. Bioeng. 1996, 49(5), 481–598.

25. Tutt, M.; Kikas, T.; Olt, J. Comparison of different pre-treatment methods on degradation of rye straw. Proceedings of 11th international conference, engineering for rural development, Jelgava, Latvia University of Agriculture Faculty of Engineering, 2012.

26. Kargbo, F. R.; Xing, J.; Zhang, Y. Property analysis and pre-treatment of rice straw for energy use in grain drying: A review. Agric. Biol. J. N. Am., 2010, 1(3), 195–200.

27. Zimbardi, F.; Viggiano, D.; Nanna, F.; Demichele, M.; Cuna, D.; Cardinale, G. Steam Explosion of Straw in Batch and Continuous Systems. Appl. Biochem. Biotechnol. 1999, 77–79, 117–125.

28. Junginger, M. Role of biomass in meeting future energy demands. Workshop Biomass supply challenges – how to meet biomass demand by 2020, Rotterdam, Netherlands [Online] 2012. http://www.biofuelstp.eu/biomass-workshop/pdf/junginger.pdf (accessed March 22, 2016).

29. Harmsen, P.; Hujigen, W.; Bermudez, L.; Bakker, R. Literature review of physical and chemical pre-treatment processes for lignocellulosic agro residuals. Energy research center of the Netherlands ECN (Online) 2010. https://www.ecn.nl/docs/library/report/2010/e10013.pdf (accessed May 1, 2016).

30. Kahr, H.; Alexander Jäger, A.; Lanzerstorfer, C. Bioethanol Production from Steam Explosion Pretreated Straw, Bioethanol. Ed. Lima, M.; Natalense, A. INTECH [Online] 2012. http://www.intechopen.com/books/bioethanol/bioethanol-production-from-steam-explosion-pretreated-straw (accessed May 7, 2012).

31. Verardi, A.; De Bari, I; Ricca, E.; Calabrò, V. Hydrolysis of Lignocellulosic Agro residuals: Current Status of Processes and Technologies and Future Perspectives, Bioethanol, Ed. Lima, M.; Natalense, A. INTECH [Online] 2012. http://www.intechopen.com/books/bioethanol/hydrolysis-of-lignocellulosic-biomass-current-status-of-processes-and-technologies-and-future-perspe (accessed March 12, 2016).

32. Tabil, L.; Adapa, P.; Kashaninejad, M. Agro residuals Feedstock Pre-Processing – Part 1: Pre-treatment, Biofuel's Engineering Process Technology, Ed. Bernardes, M., INTECH, 2011.

33. Badiei, M.; Asimb, N.; Jahima, J. M.; Sopian, K. Comparison of Chemical Pre-treatment Methods for Cellulosic Agro residuals. APCBEE Procedia 2014, 9, 170–174.

34. Chen, H.; Liu, Z. Multilevel composition fractionation process for high-value utilization of wheat straw cellulose. [Online] 2014, 137(7), 1–12. http://biotechnologyforbiofuels.biomedcentral.com/articles/10.1186/s13068-014-0137-3 (accessed May 8, 2016).

35. Sparkling, C.; Cone, J. W.; Pelican, W.; Hendriks, W. H. Utilization of Rice Straw and Different Treatments to Improve Its Feed Value for Ruminants: A Review. Asian-Aust. J. Anim. Sci. 2010, 23(5), 680–692.

36. Garcia-Cuero, M. T.; Gonzalez-Benito, G.; Indacoechea, I.; Coca, M.; Bolado, S. Effect of ozonolysis pre-treatment on enzymatic digestibility of wheat and rye straw. Bioresource Technol. 2009, 100, 1608–1613.

37. Gould, J. M. Alkaline peroxide delignification of agricultural residues to enhance enzymatic scarification. Biotechnol. Bioeng. 1984, 26, 46–52.

38. Anwara, Z.; Gulfrazb, M.; Irshad, M. Agro-industrial lignocellulosic agro residuals a key to unlock the future bio-energy: A brief review. J. Radia. Res. Appl. Sci. 2014, 7(2), 163–173.

39. Sonnenberg, A. Fungi can replace chemical pre-treatment of wood. https://www.wageningenur.nl/en/show/Fungi-can-replace-chemical-pre-treatment-of-wood-1.htm

40. Lu, Z. Chemical coupling in wood-polymer composites. PhD Theses, Louisiana State University and Agricultural and Mechanical College USA, 2003.

41. Prithivirajan, R.; Jayabal, S.; Bharathiraja, G. Bio-based composites from waste agricultural residues: mechanical and morphological properties. Cellulose Chem. Technol. 2015, 49(1), 65–68.

42. Ibrahim, H. A. Pre-treatment of straw for bioethanol production. Energy Procedia 2012, 14, 542–551.

43. Xue, F.; Li J.; Li G.; Yu, Z.; Huang J. Production of fuel ethanol from five species of forages under different pre-treatment conditions. Guizhou Nongye Kexue, 2009, 37(1), 111–115.

44. Bhagwat, S.; Ratnaparkhe, R.; Kumar, A. Biomass Pre-treatment Methods and Their Economic Viability for Efficient Production of Biofuel. Br. Biotechnol. J. [Online] 2015, 8(2), 1–17. www.sciencedomain.org (accessed May 9, 2016).

45. Peng, B. L.; Dhar, N.; Liu, H. L.; Tam, K. C. Chemistry and Applications of Nanocrystalline Cellulose and Its Derivatives, A Nanotechnology Perspective. Can. J. Chem. Eng. 2011, 89(5), 1191–1206.

46. Siro, I.; Plackett, D. Micro fibrillated cellulose and new nanocomposite materials, A review. Cellul. 2010, 17, 459–494.

47. Rebouillat, S.; Pla, F. State of the Art Manufacturing and Engineering of Nanocellulose: A Review of Available Data and Industrial Applications. J. Biomater. Nanobiotechnol. 2013, 4(2), 165–188.

48. Yokoyama, T.; Huang, C. C. Nanoparticle Technology for the Production of Functional Materials. Powder Part. 2004, 23, 7–17.

49. Miroslawa, S.; Janusz. K.; Antczak Tadeuez. A. Nanotechnology – Methods of Manufacturing Cellulose Nanofibers. Fibers Tex. E.EU. 2012, 20, 2(91), 8–12.

50. Patel, B. H. Nano-Particles and Their Uses in Textiles. Indian J. T. 2007.

51. Moon, R. J.; Martini, A.; Nairn, J.; Simonsen, J.; Youngblood, J. Cellulose nanomaterial review, structure, properties and nanocomposites. Chem. Soc. Rev. 2011, 40, 3941–3994.

52. Kalia, S.; Dufresne, A.; Cherian, B. M.; Kaith, B. S.; Avérous, L.; Njuguna, J.; Nassiopoulos, E. Cellulose-Based Bio- and Nanocomposites, A Review. Int. J. Polym. Sci. [Online] 2011, Vol. 2011, 1–35 http://www.hindawi.com/journals/ijps/2011/837875/ (accessed May 8, 2016).

53. Frone, A. N.; Panaitescu, D. P.; Donescu, D. Some aspects concerning the isolation of cellulose micro- and nano- fibers. U. P. B. Sci. Bull., Series B. 2011, 73(2), 133–152.

54. Henriksson M.; Henriksson G.; Berglund L. A.; Lindstrom, T. An environmentally friendly method for enzymeassisted preparation of microfibrillated cellulose (MFC) nanofibers. Eur. Polym. J. 2007, 43(8), 3434–3441.

55. Paakko M.; Ankerfors M.; Kosonen H.; Nykanen A.; Ahola S.; Osterberg M.; Ruokolainen J.; Laine J.; Larsson P. T.; Ikkala O.; Lindstrom T. Enzymatic hydrolysis combined with mechanical shearing and high-pressure homogenization for nanoscale cellulose fibrils and strong gels. Biomacromolecules 2007, 8(6), 1934–1941.

56. Lee S. Y.; Chun S. J.; Kang I. A.; Park J. Y. Preparation of cellulose nanofibrils by high pressure homogenizer and cellulose-based composite films. J. Ind. Eng. Chem. 2009, 15(1), 50–55.

57. Yadav, T. P.; Yadav, R. M.; Singh, D. P. Mechanical Milling, a Top Down Approach for the Synthesis of Nanomaterials and Nanocomposites. Nanosci. Nanotechnol. 2012, 2(3), 22–48.

58. Stark, N. M.; Rowlands, R. E. Effects of Wood Fiber Characteristics on Mechanical Properties of Wood/Polypropylene Composites. Wood Fiber Sci. 2003, 35(2), 167–174.

59. Pfister, D. Green composites and coatings from agricultural feedstocks, PhD Theses, Iowa State University, Ames, Iowa, USA, 2010.

60. Yam, K. L.; Gogoi, B. K.; Lai, C.; Selke, S. Composites from compounding wood fibers with recycled high density polyethylene. Polym. Eng. Sci. J. 2004, 30(11), 8, 693–699.

61. Amuthakkannan, P.; Manikandan, V.; Winowlin Jappes, J. T.; Uthayakumar, M. Effect of Fiber Length and Fiber Content on Mechanical Properties of Short Basalt Fiber Reinforced Polymer Matrix Composites. Mater. Phys. Mech. 2013, 16, 107–117.

62. Shao-Yun Fu, S.; Lauke, B. Effect of fiber length and fiber orientation distribution on the tensile strength of short – fiber reinforced polymers. Compos. Sci. Technol. 1996, 56(10), 1179–1190.

63. Samir, M.; Alloin, F.; Gorecki, W.; Sanchez, J. Y.; Dufresne, A. Nanocomposite materials reinforced with cellulose whiskers in atactic polypropylene, effect of surface and dispersion characteristics. Macromolecules 2004, 108(30), 4839–4844.

64. Wittaya, T. Microcomposites of rice starch film reinforced with microcrystalline cellulose from palm pressed fiber. Int. Food Res. J. 2009, 16, 493–500.

65. Wu, Q.; Henriksson, M.; Liu, X.; Berglund, L. A. A High Strength Nanocomposite Based on Microcrystalline Cellulose, Polyurethane. Am. Chem. Soc. 2007, 8(12), 3687–3692.

66. Silva Perez, D. D.; Tapin-Lingua S.; Lavalette. A.; Barbosa. T.; Gonzalez, I.; Bras, G. S. J.; Dufresne, A. Impact of micro/nanofibrillated cellulose preparation on the reinforcement properties of paper and composites films. 2010 Int. Conference on Nanotechnology for the Forest Products Industry. Dipoli congress center Espoo, Finland, 2010.

67. Siro, I.; Plackett, D. Micro fibrillated cellulose and new nanocomposite materials, a review, Cellulose 2010, 17, 459–494.

68. Brinchi, L.; Cotanaa, F.; Fortunati, E.; Kenny, J. M. Production of monocrystalline cellulose from lignocellulosic agro residuals: Technology and applications. Carbohydrate Polym. 2013, 94, 154–169.

69. Zhu, S., Wu, Y., Yu, Z., Liao, J., Zhang, Y. Pre-treatment by microwave/alkali of rice straw and its enzymic hydrolysis. Process Biochem. [Online] 2014, 40, 3082–3086

70. El Messiry, M. Morphological Analysis of Micro-fibrillated Cellulose from Different Raw Materials for Fiber Plastic Composites. J. Tex. Sci. Eng. [Online] 2014, 4(5), 1–7 http://www.omicsgroup.org/journals/morphological-analysis-of-microfibrillated-cellulose-from-different-raw-materials-for-fiber-plastic-composites-2165-8064.1000166.php?aid=30266 (accessed May 7, 2016).

71. Abba, H.; Nur I.; S. Salit, S. Review of Agro Waste Plastic Composites Production. J. Minerals Mater. Charact. Eng. 2013, 1(5), 271–279.

72. Hietala, M. Extrusion Processing of Wood-Based Biocomposites. PhD. Thesis, University of Oulu, Finland, 2013.

73. Lewandowski, K.; Zajchowski, S.; Mirowski, J.; Kościuszko, A. Studies of processing properties of PVC/wood composites. Chemk. 2011, 4, 333–336.

74. Gardner, D. J. Extrusion of Wood Plastic Composites. [Online] http://www.entwoodllc.com/PDF/Extrusion%20Paper%2010–11–02.pdf (accessed May 15, 2015).

75. Farsi, M. Chapter 10, Thermoplastic Matrix Reinforced with Natural Fibers: A Study on Interfacial Behavior, Some Critical Issues for Injection Molding [Online], Ed. Wang J. INTECH, 2012. http://www.intechopen.com/ (accessed May 15, 2015).

76. Material safety data sheet wood flour. [Online] http://www.rotdoctor.com/products/msdspdf/Wood_Flour.pdf (accessed May 15, 2015).

77. Lesslhumer, J. Extrusion of wood polymer composites, (presentation) Int. Conf. Challenges in extrusion of polymer. Leoben, Austria. [Online] 2014 http://www.polyregion.org/files/attachments/13135/449166_04_Lelhumer_presentation_Extrusion_of_WPC_Lelhumer_2014.pdf (accessed May 15, 2015).

78. Hunnicutt, B. Injection Molding Wood-Plastic Composites. [Online] http://www.ptonline.com/articles/injection-molding-wood-plastic-composites (accessed May 15, 2015).

79. Peter Lee, Blow Molding Behaviour of Wood Fiber Reinforced Polypropylene Composites. MSc. Thesis, Graduate Department of Forestry, University of Toronto, 1997.

80. Renard, C. Injection molding WPC. Bachelor Thesis in design and product realization Stockholm, Sweden, 2011.

81. Han, S. O.; Karevan, M.; Na Sim, I.; Bhuiyan, M. A. Understanding the Reinforcing Mechanisms in Kenaf Fiber/PLA and Kenaf Fiber/PP Composites: A Comparative Study. Int. J. Polym. Sci. [Online] 2012, 1–8, http://www.hindawi.com/journals/ijps/2012/679252/ (accessed May 5, 2016).

82. Wolcott, M. P.; Englund, K. A. Technology Review of Wood-Plastic Composites. 33rd International particleboard/composite materials symposium 1999, 103–111. Washington State University, USA.

83. Injection Molding. [Online] www.m-ep.co.jp/en/pdf/product/novaduran/molding.pdf (accessed May 5, 2015).

84. Yuan, Q.; Wu, D.; Gotama, J.; Bateman, S. Wood Fiber Reinforced Polyethylene and Polypropylene Composites with High Modulus and Impact Strength. J. Thermoplast. Compos. Mater. 2008, 21(3), 195–208.

85. Yasina, M.; Bhuttob, A. W.; Bazmia, A. A.; Karimb, S. Efficient Utilization of Rice-wheat Straw to Produce Value-added Composite Products. Int. J. Chem. Environ. Eng. 2010, 2(2), 136–143.

86. Chard, J. M.; Creech, G.; Jesson, D. A.; Smith, P. A. Green composites: sustainability and mechanical performance. Proceeding of 18th international conference on composite materials ICC 18, Korea, 2011.

87. Elkamel, A.; Fatoni, R.; Simon, L. Optimal Product Design of Wheat Straw Polypropylene Composites. Proceedings of the 3rd International Conference on Industrial Engineering and Operations Management. Istanbul, Turkey, 2012, 2116–2126.

88. Viola, N. M.; Battistelle, R. A. G.; Valarelli, I. D.; Casarin, S. A. Use of waste plastic and wood flour in the production of composite. [Online] https://www.pomsmeetings.org/confpapers/043/043–1348.pdf (accessed May 8, 2016).

89. Abba, H. A.; Nur, I. Z.; Salit, S. M. Review of Agro Waste Plastic Composites Production, Journal of Minerals and Materials Characterization and Engineering 2013, 1, 271–279.

90. Běhálek, L.; Lenfeld, P.; Seidl, M.; Bobek, J.; Ausperger, A. Friction properties of composites with natural fibers, synthetic and biodegradable polymer matrix. NANCON, 2012, Brno, Czech Republic, EU, 2012.

PART IV

COMPOSITE MATERIAL TESTING METHODS: FIBER, YARN, FABRIC, POLYMER AND COMPOSITE

CHAPTER 8

TESTING METHODS FOR COMPOSITE MATERIALS

CONTENTS

8.1   INTRODUCTION

The composite material is a bonding of several materials, which are of different physical and mechanical properties, to create a new material with properties that cannot be achieved by any of the components acting alone in the final form of the designed part. The performance of the composite depends of the properties of each component and how strongly they bonded to each other. Otherwise, the properties of the composite will be that of the weakest one as in the chain strength is that the minimum link. The heterogeneity of the properties of composite components requires uniform distribution of the components in space of the composite architect and minimizes the anisotropic properties of the components used. This required continues testing of each component with the main objectives:

1.   Check the properties of the raw material used;
2.   Check the designed composite properties;
3.   Quality assurance of the produced composite properties;
4.   Compliance of composite material to applicable standards and specifications;
5.   Verify supplier specifications.

8.2   TESTING OF THE DIFFERENT COMPOSITE'S COMPONENTS

In the natural fiber polymer composite, the main components are fibrous material and the matrix. Each should be tested for several properties related to the required performance of the final composite before it is used in manufacturing process.

8.2.1   SAMPLING OF TEST SPECIMEN

The textile fibers and textile material, in general, are heterogeneous material. Choice of the sample should represent the population in order to insure the confidence of the results of the test. Random sample, numerical sample or biased sample may be chosen. For each test of the fibers, there is a procedure of the sampling to verify the required confidence of the results. ASTM D1441 – 12, covers the sampling of cotton fibers, while ASTM D2258 – 99(2012) for sampling of yarn, for fabric each type of test its sampling procedure as stated in the ASTM standards of the test such as ASTM D737–96 for air permeability, ASTM D3776/D3776M – 09a (2013) for mass per unit area (Weight) of fabric, ASTM D5034 for fabric breaking strength and elongation.

Determining sample size is a very important issue because samples that are too large may waste time, resources and money, while samples that are too small may lead to inaccurate results. In many cases, we can easily determine the minimum sample size needed to estimate a process parameter, such as the population mean μ. To prepare an unbiased sample in which the population of the different parameters in the sample are presenting as those in the bulk or through sampling each product has equal chance of being chosen when withdraw the sample. The size of samples is governed by nature of the test, degree of accuracy expected in the results, variability expected in the material. If the measured parameter of the tested phenomena follows a normal distribution, assuming an error E% and standard deviation σ, for the certain value of confidence t the number of the samples required is:

$$n = (t.\sigma/E)^2 \qquad (1)$$

For confidence limit 90%, the value of t is equal to 1.645, for 95%, t is equal to 1.96, while for 99%, t is equal to 2.58. The accepted value of error depends on the importance of the tested value.

For example, if fabric has a mean strength 800 N and standard deviation $\sigma = 5$ N, number of samples to be tested with error 5% and with confidence limit 95% will be $n = 3$ samples.

8.2.2 FIBROUS REINFORCEMENT

The components of material used for natural fibers composites are the natural fibers, chemical coupling agent and resin. Each component is tested for different properties as well as the final composite material to comply the target properties. The reinforcement may be fiber, yarns, 2-D and 3-D woven fabrics, knitted fabric, braided fabric or nonwoven mat. Each material requires number of different testing procedures. The essential tested properties of the fibrous reinforcement may be fibers, yarns, and fabrics.

8.2.2.1 Fibers

The main prop erties of the natural fibers may be classified as shown in Figure 8.1.

8.2.2.2 Yarns

The parallel yarn reinforcement can be used in forming composites, made of natural fibers, or used to form 2-D or 3-D structure fabrics. The main yarn properties are shown in Figure 8.2.

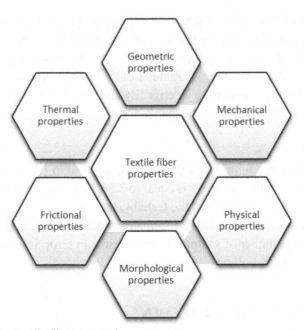

FIGURE 8.1   Textile fiber's properties.

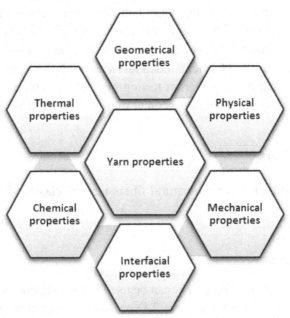

FIGURE 8.2   Yarn properties.

8.2.2.3 Fabrics

When fabrics are used as reinforcement, several properties should be tested to insure it suitability to give the required performance of the composite. The main fabric properties are shown in Figure 8.3.

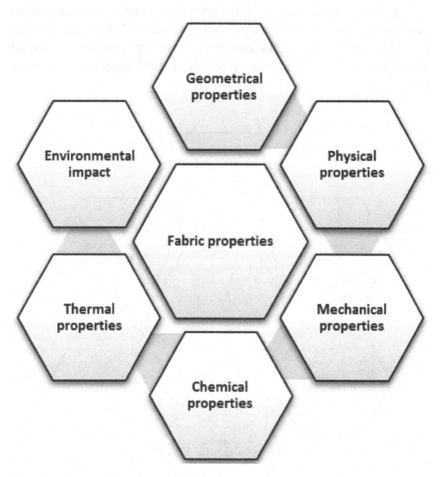

FIGURE 8.3   Textile fabric properties.

8.3   POLYMER PROPERTIES TESTING

The polymers are used to form the matrix, which have different properties depending on the length of the polymer chain and chain structure. There are thermoset and thermoplastic polymers. When the thermoplastic polymers are heated, the de-bonding between chains occurs and the viscosity of the polymer increases, on the contrary to thermoset polymers – on heating of the polymer no movement between the molecules occurs.

The density, mechanical properties (tension, compression, shear), glass transition temperature "T_g," melting temperature "T_m," thermal expansion properties, resistivity, coefficient of thermal expansion, fiber polymer interfacial strength are important properties to be determined, as given in Figure 8.4.

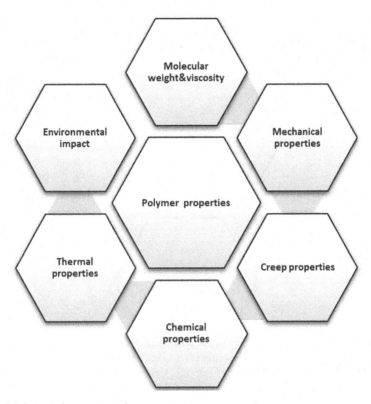

FIGURE 8.4   Polymer properties.

TABLE 8.1 Essential Tests for Composites Components

I. Reinforcement

a. Fibers

1. Physical	Diameter, Crimp, Friction	Density, Moisture Absorption	Thermal Properties	Electrical Properties
2. Mechanical	Tensile Properties Single Fiber, bundle fibers	Creep	Torsion	

b. Yarns

1. Physical	Count, Diameter, Twist, Number of Ply, Friction	Density		
2. Mechanical	Tensile Strength, Creep	Interfacial Shear Strength	Ballistic	Fatigue

c. Fabric

1. Physical	Fabric Structure, Areal Density, Fiber Alignment, Porosity, Air Permeability	Density, Moisture Absorption	Thermal Properties, Flammability	Electrical Properties
2. Mechanical	Tensile Strength, Creep	Interfacial Shear Strength	Fatigue	Ballistic

II. Matrix

a. Polymer

1. Physical	Density, Moisture Absorption	Thermal Analysis, Melting Temperature T_m Glass Transition Temperature T_g	Thermal Expansion Properties, Coefficient of Thermal Expansion	Electrical Properties
2. Mechanical	Tensile Strength, Shear Strength	Compression Strength	Creep	Ballistic

TABLE 8.1 (Continued).

b. Resin

1. Physical	Density, Moisture Absorption	Resin Extraction	Resin Flow, Gel Time	Surface Tack, Drape
2. Mechanical Adhesives	Tensile Strength	Shear Strength	Bonding Joint Strength	
III. Composites				
1. Physical	Density	Fiber, Matrix, Voids Percentage	Fiber Volume Fraction	Coefficient of Thermal Expansion
2. Mechanical	Strength, Strain, Elastic Modules, Passion's Ratio	Adhesives	Shear: In-Plane Shear Stiffness: Out-of-Plane (Inter-Laminar) Shear	Compression, Drape, Creep, Impact, Fatigue Strength
	Bearing		Impact Strength Compression Strength After Impact	Fracture Mechanics Extension, Shear, Tear, Double Cantilever Beam Flexure, End–Notched Flexure, Mixed Bending and Tension, Bending And Compression
3. Composite Defects By Non-Destructive Methods	Visual Inspection Thermography	Tap Testing Shearography	Ultrasonic	X-Ray

8.4 COMPOSITE ESSENTIAL TESTING

According to the required performance of the final composite design, the type of testing methods is selected. The classification of the type of tests for composite material are: component identification, physical properties, mechanical properties, adhesives properties, composite's defects. Table 8.1 gives summery of the essential tests required.

The number of the tests to determine the material properties and the final composite properties depends on the end use of the composite material and the type of loading and finally, the environment surround the composite. Consequently, not all the tests should be carried out and the designers should decide which value of the properties should be tested. In the next paragraphs we will give some overall pictures about the methods of the testing and the standards used. Ambient conditions for all testing must be carried out under constant standard atmospheric conditions. The standard atmosphere for textile testing involves a temperature of $20\pm2°C$, and $65\pm2\%$ rh. The samples must be pre-conditioned at least 24 hours under constant standard atmospheric conditions to attain the moisture equilibrium.

8.5 REINFORCEMENT TESTING METHODS

8.5.1 FIBER TESTING

Due to the variability of fiber types, the different testing methods and procedures are used to define the physical and mechanical properties. Table 8.2 demonstrates the principles of some of those testing methods.

TABLE 8.2 Fiber Testing Methods

Fiber properties	Principles	ASTM	Reference
Fiber Length			
Optical measurement	Fibrograph measurements provide a relatively fast method for determining the fiber length and uniformity of the fibers in a sample of cotton in a reproducible manner. The digital fibrograph uses an optical sensor to scan the fiber beard and draw the relation between the number of the fibers scanned in the beard against the fiber length, in order to get designation of the length and values reported, such as 2.5% span length, 50% span length, and uniformity ratio: (50 ML/2.5SL). There are several apparatus for the determination of fiber length, such as HVI, AFIS. The parameters of fiber length are calculated as illustrated in Figure 8.5. OA = Mean Length (ML), OB = Upper Half Mean Length (UHML), Uniformity Ratio (UR) = (LL1/L2L3) × 100, Uniformity Index (UI) = (MI/UHML) × 100 The value of UR is in the range of 45–50% while the value of UI is normally between 70 and 80%.		
Capacitive method	For wool fiber capacitive method is used to draw the cumulative frequency diagram (Automatic version of WIRA fiber diagram "ALMETER"). The wool fiber cumulative frequency diagram is measured. Hence the mean fiber length of the wool can be calculated, take 10% interval length between 5% and 95% of the total fiber length.		

Fiber Diameter

Gravitational method	The fiber fineness is an expression of the fiber diameter, however the cross section shape of natural fiber is not constant along the fiber itself and varied in the shape too. So it is preferable to be expressed in weight per unit length mg/cm or, what we call it "count," which is expressed in direct or indirect systems (Ne, Nm, tex or denier). To define fiber diameter, assuming fiber density ρ and the weight per unit length "m" g/m, then the fiber diameter can be calculated as: $$d = (4\rho(m/\pi))^{0.5}$$	[1]
LASERSCAN	Fiber snippets are dispersed into isopropanol water mixture and the resultant suspension then flow through a measurement cell. As they pass through the cell, the fiber snippets intersect a thin beam of light generated by a laser. This beam is directed at a measurement detector. The detector produces an electrical signal that is proportional to the amount of light incident upon it, which is proportional to the fiber diameter.	
Optical Fiber Diameter Analyzer (OFDA)	Magnifies and captures images of the individual fibers using a video camera and analyzed almost instantaneously by computer. Fibers and specially wool types. The OFDA instrument is based on automatic image analysis technology and was recently introduced to provide a rapid, accurate measurement of average fiber diameter (AFD) and diameter distribution (SD) of textile fibers for measuring fiber diameter parameters of sheep and goats.	[2, 3]
Scanning Electron Microscope (SEM)	The individual fibers were identified in the images, their diameters measured, cumulative probability distributions plotted, and histograms of the number of fibers within selected class interval of fiber diameters can be prepared.	

TABLE 8.2 (Continued).

Fiber properties	Principles	ASTM	Reference
Microprojection	The diameter and distribution values have been determined using the Projection Microscope. Projected fiber images are individually measured at 500X magnification.	D2130–13	[4]
Ensemble Laser Diffraction Technique	ELD works by passing a collimated laser beam through a group of fibers and measuring the scattering of the transmitted light. The radial scattering profile is directly related to the diameter distribution of the fibers.		[5]
Vibroscope	Manmade fibers have regular cross section. Gravimetric method can be used or the optical microscope method. Also Vibration method is usually used for the determination of the fineness of individual fiber, especially manmade fiber, where the fiber is clamped at one end and led over a knife edge support loaded by 'W,' Figure 8.6 and is induced a vibration of 'f' frequency at the other end, as illustrated. The frequency of the vibration increased till reaching its natural frequency. $$M = 0.25T \times (1/l \times f_1)^2$$ where, f_1 = natural fundamental frequency of vibration (c/s); T = tension; M = fiber mass per unit length (gm/cm); L = free fiber length =l/2; l = wave length; Tension range = 0.3 to 0.5 cN/tex, usually applied by weighted clip on the end of the fiber. The frequency, which given maximum vibration amplitude, is the fiber resonance frequency, from which the linear density is measured.		
Fiber Strength			
Single fiber strength	Several instrumentations are used to measure the fiber strength under constant rate of extension or loading. The fiber is fixed in the gripper, Figure 8.7, the gage length will depend on the type of fiber gauge length. For instance, for synthetic fiber 150 mm.	– C1557–14 – D3379	

Bundle fiber strength	The single fiber strength and the bundle strength were found to be highly correlated. Natural fibers are tested at zero or 3.175 mm gauge. Stelometer, Pressley and HVI. Tensile test for carbon graphite yarns roving tow, for specimens, set the distance between the grips to 150 mm. In all tests, all specimens should be conditioned for 24 h and tested at 23±2°C and 50±10% relative humidity, then no testing for moisture adsorption is required.	– D1445M – 12 – D1294 – 05(2013) – D5867 – Tow tests D 4018
Fiber-matrix interfacial bond	Several testing method can be used to characterize the fiber-matrix interfacial (IFSS) behavior. The various number of direct micromechanical methods have been developed to determine the interface shear strength. Figure 8.8 illustrates the different methods, one is suitable for the testing of certain application while the other for different ones. It should be mentioned that the principal problem usually is the interfacial adhesion between the natural fiber and the matrix that governs the final properties of a composite. The value of the (IFSS) $= P/(\pi\, d_f)$, where, P – pulling force, d_f = fiber diameter	STP452 [6–14]
Single fiber pull-out test	In the single fiber pull-out test, single fiber is embedded in the film of polymer and the other end fixed in the jaw of the testing machine, the fiber is pulled out from the polymer and the interfacial shear stress can be calculated for the unbroken fibers.	
Single fiber fragmentation test	Single fiber embedded in specimen of matrix material is tested on tensile. The fiber will break in different places along its length and continue to fragmentize with the increase of strain. Test is completed when the fiber is no more francized.	
Micro bond test	The micro bond test is one of the widely used for single fiber-matrix interfacial bond to determine the IFSS.	

TABLE 8.2 (Continued).

Fiber properties	Principles	ASTM	Reference
The push-out test	In this test, which is a reverse of the pull out test, flat tip indenter is lowered onto the fiber, touching the fiber and beginning applying a load. Bonds start to break between the matrix and the fiber and the fiber begins to slide out of the matrix. The bond force between the matrix and fiber is recorded.		
Moisture absorption	Moisture regain of the fibers is determined. The test includes oven drying the fibers at 105°C for 4 hours and then allowing the sample to absorb water under the standard testing conditions for 24 hours. The percent regain is calculated as the ratio of the amount of water absorbed to the dry weight of sample.	–D2495	

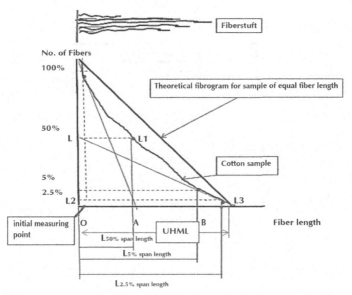

FIGURE 8.5 Fibrograph.

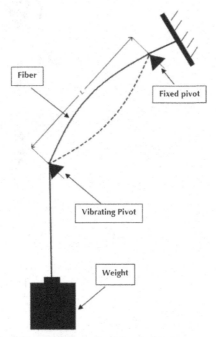

FIGURE 8.6 Sketch of the measuring principle of vibroscope.

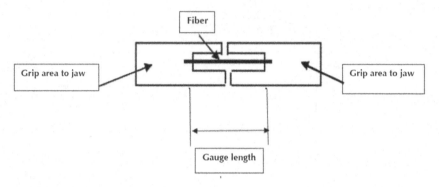

FIGURE 8.7   Fiber clamp.

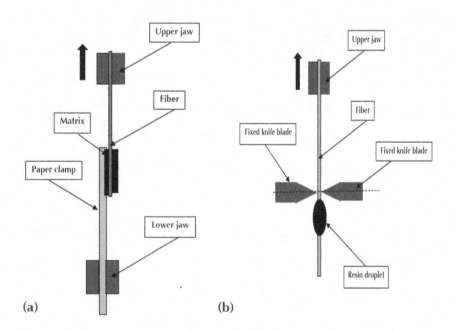

(a)   (b)

FIGURE 8.8   *(Continued)*

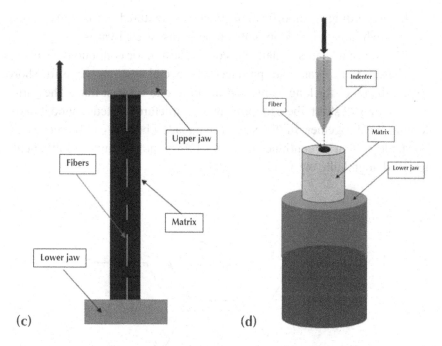

(c)

(d)

FIGURE 8.8 Method of measuring fiber-matrix interfacial force. (a) Single fiber pull-out test; (b) Micro bond test; (c) Single fiber fragmentation test; (d) Push out test.

8.5.2 Yarns Testing Methods

8.5.2.1 Introduction

The yarn is main element in the construction of woven fabric. Its physical and mechanical properties will define the end use of the fabrics. As illustrated in Figure 8.9, Yarns can be classified to:

1. According to fiber length staple yarns, continuous mono filament or multi filament yarns.
2. According to how many stages of twist applied on the fibers: single, ply, cord or cabled.
3. According to the decorative features added to the yarn "novelty yarns," such as slub, flack, flake, spiral, boucle, knob, knot, spot, chenille, tweed yarns.

4. According to modification processes, textured yarns, bulk yarns, high bulk yarns, loop bulk, air jet yarns, stretch yarns.

5. Core spun yarns: usually the core is elastomeric continuous filament.

These types of yarns are produced on several systems – one for short staple, the other for long staple and the third one for bast fiber. The yarns are assembly of short fibers to form endless structure twisted around its axis to bind them together. In the case of manmade fibers, it can be processed on staple system or continues filaments (As monofilament, multifilament, textured, high bulk, etc.).

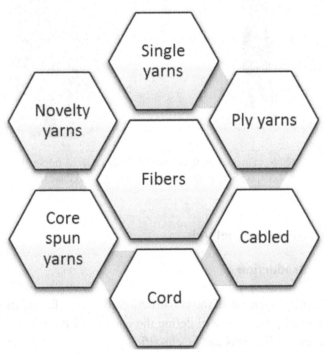

FIGURE 8.9   Types of yarns.

8.5.2.1.1 Single Yarn

Single yarn produced by spinning of staple fibers on one of the different spinning systems (ring spinning, compact spinning, open end spinning, jet spinning, woolen spinning, worsted spinning, bast fiber spinning) using twist as a binder of the staple fibers together to give the yarn its strength and integrity.

8.5.2.1.2 Ply Yarn

In order to increase the strength of the yarns, two or more single yarns are twisted together on twisting machine. Second twisting process is shown in Figure 8.10. Direction of twist z/s or s/z, in some cases z/z.

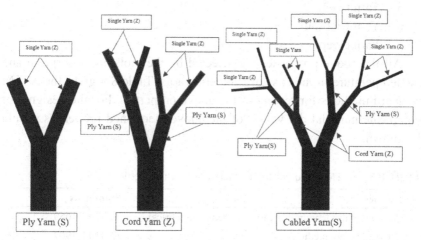

FIGURE 8.10   Sketch of different yarn structures.

8.5.2.1.3 Cord Yarn

Stronger yarns are produced by twisting of ply yarns through third stage twisting process. Direction of twist z/s/z.

8.5.2.1.4 Cabled Yarn

Several cord yarns are twisted together forming more strong yarn. Fourth twisting stage. Direction of twist z/s/z/s.

8.5.2.2 Yarn Properties

Whichever the yarns are used as reinforcement or to form the fabric to reinforce the composite, their properties are of primary importance to the mechanical properties of the composite. The main properties of the yarn are:

1. Yarn count;
2. Yarn diameter;
3. Yarn unevenness;
4. Yarn mechanical properties;
5. Yarn twist;
6. Yarn friction;
7. Yarn creep.

All the above properties are affected by the moisture absorption and the temperature in the case of synthetic yarns. Table 8.3 gives the ASTM standard methods for testing of the above properties. For the designer of composite material the values of the above mentioned properties should be known.

TABLE 8.3   ASTM Standards for the Measuring Yarn Properties

Properties	Standards
Yarn Count (short lengths-for yarns raveled from fabrics)	ASTM D1059
Yarn Count (skein method)	ASTM D1907
Yarn, Sliver, or Roving – Unevenness and Hairiness (Uster Tester 5)	ASTM D1425
Yarn Coefficient of Friction (yarn to solid object)	ASTM D3108
Skein Shrinkage (hot air or boiling water)	ASTM D2259
Yarn Tensile Properties (single end)	ASTM D2256
Twist in Single Yarn	ASTM D1422
Twist in Plied Yarn	ASTM D1423
Yarn creep	ASTMD5262–07(2012)

8.5.2.3 Yarn Count

There are two systems to define the material count:
1. **Indirect system**: (Fixed–length system): defined as number of unit length per unit mass. The higher the count number, the finer the yarn. The examples are:
   - English count (Ne) = (Length (yd.)/840)/weight (pound)
   - Linen count = (Length (yd.)/300)/weight (pound)
   - Metric count (Nm) = Length (km)/weight (kg)
   - Worsted count = (Length (yd.)/560)/weight (pound)
2. **Direct system** (Fixed-weight system): Number of mass per unit length. The higher the count number, the coarser the yarn.
   - *tex*: weight (g) per length of 1000 (m).
   - Denier: weight (g) per length of 9000 (m).

Usually, the unit length is varied for each system as well as the weight units. For cotton count, the unit length is 840 yd. and unit weight one pound while for metric count, the unit length is one meter and unit weight one gram. To convert from one system to another use the constants given in Table 8.4.

TABLE 8.4   Conversion Table for Yarn Counts

Count	*tex*	Ne	*den*	Nm
tex	1	590.54/Ne	*den*/9	1000/Nm
Ne	590.54/*tex*	1	5314.9/*den*	Nm × 0.5905
den	*tex* × 9	5310/Ne	1	9000/Nm
Nm	1000/*tex*	1.6806 × Ne	9000/*den*	1

There are several equations can be used for the yarn diameter (cm) calculations, Peirce equation:

$$d = 0.0907/(Ne)^{0.5} = 0.0036(tex)^{0.5} \qquad (2)$$

This equation is modified to take into consideration the different spinning systems:

$$d = 0.00357(tex/(\phi \cdot \rho_f))^{0.5} \qquad (3)$$

where: d – yarn diameter "cm," *tex* – yarn count, Φ – yarn packing coefficient, ρ_f – fiber density "g/cm³."

For blended yarns, the average fiber density is given by the following equation:

$$(1/\rho) = (\Sigma_1^n (\rho_i/\rho_{fi})) \tag{4}$$

where: ρ – average fiber density of the blended fibers; p_i – weight fraction of the ith component; ρ_{fi} – fiber density of the ith component; n – number of components of the blend.

One can calculate the diameter of synthetic monofilament yarn or fiber with the following formula:

$$d = (0.00014\ den/\rho)^{0.5}\ mm \tag{5}$$

where: ρ – yarn density; *den* – yarn count in denier.

The yarn diameter is a function of the yarn count. Figure 8.11 gives the change of the yarn diameter for different systems of expressing the yarn count Ne, Nm, *tex*, *den*.

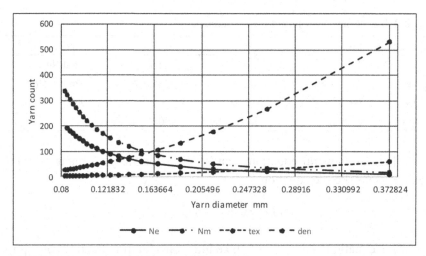

FIGURE 8.11   Yarn count versus yarn diameter.

Yarns of the same material but with different diameter have different mechanical properties, such as yarn tenacity, bending stiffness, creep properties, resilience, bulkiness, thermal properties, water absorbance, and porosity. One of the most influencing parameter is the yarn surface area which controls a number of properties, like frictional properties between fibers, size of the pores between fibers in the yarn structure, packing density, the yarn-matrix bonding, while the cross section area affects most of yarn mechanical properties. Figure 8.12 shows the relation between the yarn surface area and yarn cross-section area.

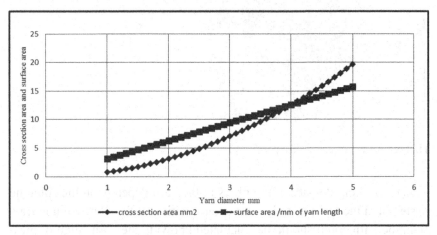

FIGURE 8.12 Effect of the yarn diameter on the surface area/cm and cross sectional area.

The fabric properties will also change, such as fabric thickness, fabric tightness, fabric weight per m², fabric thickness, fabric drape ability, moisture absorption, porosity, surface friction and comfort, etc. The mechanical properties will change too as a function of yarn diameter, like strength, tear strength, shear property.

8.5.2.4 Yarn Diameter

The yarn diameter consists of fibers packed together with different radial packing density across the radius of the yarn, as shown in Figure 8.13.

There are several formulas to calculate the diameter of the yarn which depends on the type of spinning system, hence the yarn density varied from one system to another. For example, open-end spun yarns are 20% larger in diameter than ring spun yarns. The compact spun yarns are more compact than the ring spun yarns with the same count.

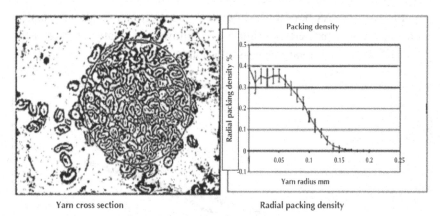

Yarn cross section   Radial packing density

FIGURE 8.13   Yarn cross section.

The packing density of the fibers in the yarn depends on the spinning system and the level of twist. For determination of the fiber volume fraction, the yarn count should be calculated knowing the yarn count or measuring the yarn diameter. Hence the fiber or yarn diameter is not constant, for practical purposes count systems are used to define diameter. The yarn density can be calculated knowing yarn diameter as:

$$\text{Yarn density (g/cm}^3) = 4\ tex\ 10^{-5}/\pi\ d^2 \tag{6}$$

8.5.2.4.1   Measurements of Yarn Diameter

Yarn diameter is an important determinant of many fabric parameters and properties, such as cover factor, porosity, thickness, air permeability, fabric appearance. The design of fabric properties for different end use depends on the yarn diameter. There are many methods based on different types of sensors used for characterization of yarn diameter which are summarized in Table 8.5. The difference between the measuring techniques is analyzed by several researchers [15–27]. However, the most common method is the optical scanning [23].

TABLE 8.5 Principles of Diameter Measurement

Method	Principles	Reference
Capacitive scanning	The yarns passed through a capacitor and the variation of its capacity is measured and analyzed to give the yarn diameter and its variation.	[23–25]
Optical scanning	Measuring yarn diameter with dual light beams perpendicular to each other. Hence the yarn diameter is not round, this method reduces shape error caused by irregular yarn cross-sections.	[23–27]
Infra-red scanning	Principle of optical measurement using infra-red light where infrared light sensor operating with a precision of 1/100 "mm" over a measuring field length of 2 mm and at a sampling interval also of 2 mm.	[19, 20]
Flying Laser spot scanning	When a yarn is passed in the scanning area, the flying spot generates a synchronization pulse that triggers the sampling. The width between the edge of the first and the last light segment determines the diameter of the yarn. Depending on the spot size and specimen feeding speed, the measurement values may vary.	[19, 26]
Photoelectric method	Beam of light is directed onto photoelectric cell through a narrow slit where the yarn is passing through. The magnitude of the current produced by photo cell is proportional to the intensity of light falling, which depends on the diameter of the yarn.	[19, 22]

The yarn cross section shape is another parameter that affects the thickness of the fabric. The yarn shape factor indicates the roundness of the yarn.

8.5.2.5 Yarn Twist

In spinning of staple fiber yarns, twist is inserted into the strand of fibers in order to hold the fibers together and impart enough yarn strength. A change in the level of twist changes the yarn strength as well as different properties, such as breaking elongation, diameter, yarn surface morphology friction properties, bending stiffness, and yarn packing density. Also, twist defines the properties of the fabric, such as fabric strength, drape ability, stiffness, resilience, surface friction and appearance of the fabric. The composite properties will be directly affected by the amount of the twit imported in the yarn.

The twist amount can be expressed by the number of turns given to the yarn per meter (tpm). The direction of the twist may be **S** or **Z** as illustrated in Figure 8.14.

Usually, the single yarns are twisted in Z direction, except for sewing threads or ply yarns. The fibers in the yarn will be inclined to its axis by angle Θ:

$$tan\Theta = \pi\,d\,(twist\,per\,cm) \tag{7}$$

Assuming the yarn density ρ is constant, then:

$$Twist\,(tpm) = k/tex^{0.5} \tag{8}$$

where: *k* is constant called twist factor.

$$k = 0.5\,tan\Theta.\,(\rho\,10^3\,/\pi)^{0.5} \tag{9}$$

$$k = (tpm)\,(tex)^{0.5} \tag{10}$$

where tpm = twist per meter.

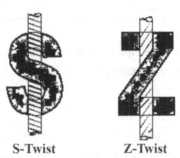

S-Twist   Z-Twist

FIGURE 8.14   Direction of yarn twist.

Thus, k is a factor relating twist level to yarn count. The derivation shows that if two yarns have the same twist factor, they will have the same surface twist angleΘ, regardless of count. Since surface twist angle is the main factor determining yarn character, then twist factor can be used to define the character of a yarn.

8.5.2.5.1 Measurement of Twist

Twist measurement is a routine test for yarns, according to ASTM D1422 for single yarns, ASTM D1423 for ply yarns. Some basic principles are discussed here.

1. *Straightened Fiber Method*: This method involves counting of the number of turns required to untwist the yarns until the surface fibers appear to be straight and parallel to yarn axis. This method is mainly used for ply and continuous filament yarns.
2. *Untwist/Twist Method*: This is the common method used for staple fiber yarns. To measure the twist level in a yarn with Z twist, the yarn is first untwisted (by a twist tester) in S direction, until the surface fibers appear to be straight and parallel to yarn axis, and continue to twist the yarn in the same direction – yarn will contract till it reaches its initial length. In this situation the number of the revelation of the yarn end will be double the original twist.

8.5.2.6 Yarn Tensile Properties

Yarn strength is the mean property that determines the fiber reinforcement of the NFPC material. In the case of the staple fibers yarns, the strength will be a function of the fiber strength, length and frictional properties, while the twist inserted for yarn formation is the predominated factor of the yarn strength. For yarns made of continuous filament, its strength depends mainly on the filament strength. Universal testing instrument, illustrated in Figure 8.15, is usually used for the tensile measurement of yarn strength. The yarn tested gauge length is 500 mm and number of samples is according to "ASTM D2256."

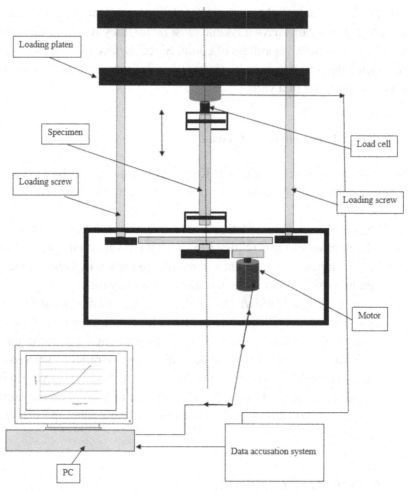

FIGURE 8.15   Sketch of the universal testing instrument.

The instruments used for determining the tensile strength are classified into three groups, based on the principle of applying the load: Constant rate of traverse (CRT); Constant rate of extension (CRE); and Constant rate of loading (CRL).

The rate of loading as determined by the "time-to-break," is an important factor, which defines the strength value recorded by means of any instrument. Lower strength recorded when the time-to-break is high and vice versa. The testing speed varied between 200 and 400 m/min depending on the material of yarn tested. The yarn stress strain curve is illustrated in Figure 8.16.

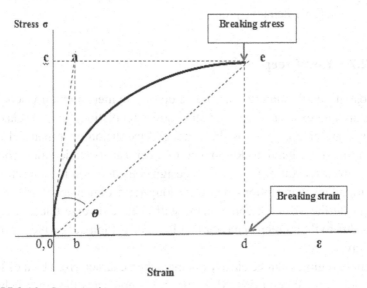

FIGURE 8.16   Stress strain curve.

From the strength elongation curve, the following values can be calculated:

1. Yielding load stress "N/*tex*"
2. Yarn tenacity stress "N/*tex*" or "MPa," Yarn tenacity = oc cN/*tex*
3. Breaking strain = od
4. Initial modulus or Young's modulus = (oa/ob) cN/*tex*
5. Work factor (2 works of rupture/ (breaking load x breaking elongation)
6. Percentage of unreturned elongation after removing load/elongation under the load
7. Fiber toughness energy absorbed before fracture (area under the load elongation curve "oed") N.cm or joule (Work of rupture).

The universal testing machine is used for the tensile or compression tests and also for the other types of tests, such as shear, flexure, peel, tears, and cyclic loading. Usually, the yarn strength is given in N/tex. In order to convert N/tex to MPa, the following equation can be used:

$$\text{Yarn strength (MPa)} = 10^3 \, \varphi \, \rho_f \, (N/tex) \tag{11}$$

where: d – Yarn diameter "cm," tex – Yarn count, φ – Yarn packing coefficient, ρ_f – Fiber density "g/cm^3."

8.5.2.7 Yarn Creep

Almost all textile materials are made up of polymers and are viscoelastic in nature and exhibit creep, as well as stress relaxation, inverse relaxation and inverse creep. Creep is the continued extension of a material under long-term static loading. Resistance to creep (or its static-strain complement, stress-relaxation) is a critical design consideration in material selection for many applications requiring long-term dimensional stability. In composite the different components will behave diverse under the same stress due to their creep behavior that has a direct impact on the formation of micro-cracks.

Three regimes can be clearly distinguished characterized by a different behavior of the creep rate [28]. Figure 8.17 represents the creep behavior which can be divided into; Zone I "primary creep," Zone II: "steady state creep," Zone III: "tertiary creep," in the last zone molecular chains start to break. High strains will start to cause necking in the yarns and will increase the local stress that further accelerates the strain until breakage. The creep may be due to tension or compression or bending loads. Several standard testing methods are used for testing the creep of the material depending on its end use (ASTM D885/D885M, Standard Test Methods for Tire Cords, Tire Cord Fabrics, and Industrial Filament Yarns Made from Manufactured Organic-Base Fibers, ASTM D 5262, Standard Test Method for Evaluating the Unconfined Tension Creep and Creep Rupture Behavior of Geosynthetics, ASTM D7337/D7337M Tensile Creep Rupture of Fiber Reinforced Polymer Matrix Composite Bars, ASTM D4762, Standard Guide for Testing Polymer Matrix Composite Materials, ASTM D6112, Standard Test Methods for Compressive and Flexural Creep and Creep-Rupture of Plastic Lumber and Shapes, ASTM D2990, Standard Test Methods for Tensile, Compressive, and Flexural Creep and Creep-Rupture of Plastics).

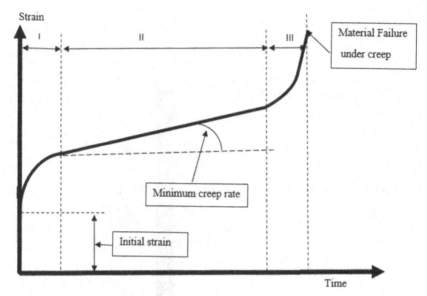

FIGURE 8.17 Creep behavior.

The designer of the composite material should include into consideration the following: creep rupture, thermal relaxation, dynamic creep under cyclic loads or cyclic temperatures, creep and rupture under multiaxial states of stress, cumulative creep effects, and effects of combined creep and fatigue. The creep test is conducted using a tensile specimen to which a constant stress is applied, often by the simple method of suspending weights. Surrounding the specimen is a thermostatically controlled furnace, the temperature being controlled by a thermocouple attached to the gauge length of the specimen, Figure 8.18.

8.5.3 FABRIC TESTING METHODS

The testing methods of the fabric may be divided into basic test to determine the fabric specifications and several other tests, depending on the end use of the fabric. The properties, related to the design of the composites, and should be tested before the manufacturing is summarized in Table 8.6.

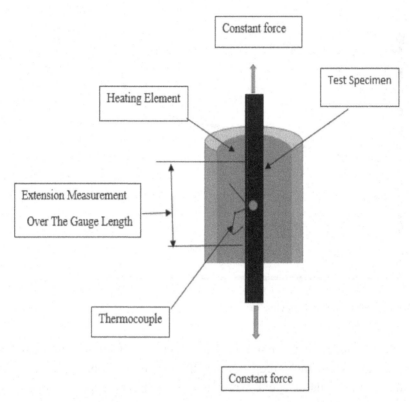

FIGURE 8.18   Principle of measuring the creep properties of the material.

8.5.3.1   Yarn Crimp

Due to the interlacing of the weft and warp, the yarns are not straight but interlocked on each other, therefore increasing the length of the yarn in each direction. The crimp increases as the number of ends and picks increases. The fabric surface topography will be affected by the crimp value which influences the fiber-matrix interfacial strength. The crimp is simply measured by raveling yarns in the direction where the crimp calculated. Crimp is calculated under a predetermined tension:

Crimp% = (Straight length of raveled yarn –
Yarn's length in fabric) × 100/Yarn's length in fabric   (12)

TABLE 8.6   Fabric Testing Standards

Properties	ASTM Standards
Physical	
Yarn crimp	D3883
Yarn count	D1059
Mass per unit area	D3776
Picks/cm and ends/cm	
Fabric thickness	D1777
Air permeability	D737
Flammability	D6413
Porosity	
Moisture content	D789
Yarn Melting point	D7138
Fabric stiffness	D1388
Mechanical	
Tearing strength	D 1424, D5587
Grab strength	D5034
Strip strength	D5035
Shear strength and Shear modulus	
Stiffness	D1388
Tear strength	D2261
Thermal	
Heat transfer	D7140
Thermal conductivity	C1696
Flammability	D1230
Electrical	
Resistivity	D257, D4496

8.5.3.2   Air Permeability

Air permeability express the resistance of fabric to the air flow passing under constant prescribed air pressure (differential between the two surfaces of a fabric) while firmly clamped in the test head, as illustrated in Figure 8.19. The value of air permeability will be equal to:

Air permeability (cm³/sec/cm²) = Rate of air flow (cm³/sec)/
Sample area (cm²)
× Pressure drop (cm water)   (13)

As it was stated before, there is a positive correlation between the air permeability and the fabric porosity. Fabric porosity, being one of the most important factors which determines the composite mechanical properties, affects the fiber volume fraction of composite and the resin flow rate.

Theoretically, permeability can be calculated by this formula:

$$K = (\mu.h.V)/\Delta P \qquad (14)$$

where: μ – fluid viscosity (pa.s); h – flow length, thickness of tested sample, (m); k – permeability coefficient; ΔP – pressure drop (pa); V – fluid speed. k is considered as a constant for the relationship between V and ΔP and is called the permeability of the fabric.

The value for k will depend on the type of porous media, the pore geometry and fluid viscosity [29].

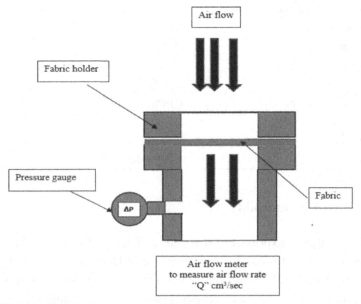

FIGURE 8.19   Sketch of the air permeability tester.

8.5.3.3   Fabric Porosity

Porosity is most important geometrical property of the pores in the fabric. The basic equation describing fluid flow in porous medium is [29]:

$$Q = K. \Delta P. A/\mu. h \qquad (15)$$

where: Q: flow rate (m³/s); A: fabric area (m²).

The porosity of the fabric used in composite as reinforcement plays an important role in manufacturing the composites as well as the final properties and the voids percentage. Volume porosity of a fabric is defined as the ratio of the volume fabric material to total fabric volume expressed as a percentage. Porosity of woven fabric is affected by some parameters like yarn structure as well as fabric specifications [30]. The total porosity of a woven fabric consists of three components: (a) the intra-fiber porosity, due to the voids within the fiber itself, (b) the inter-fiber porosity, due to the voids between the fibers, and (c) the inter yarn porosity, formed by the intersections between the yarns. The pores, within a fabric, are also influential factors in rate of flow of resin during the composite manufacturing [31]. The porosimeter's principles can be classified as:

- Liquid extrusion: extrusion flow porosimetry, extrusion porosimetry.
- Liquid intrusion: mercury intrusion porosimetry, non-mercury intrusion porosimetry.
- Gas Adsorption: vapor adsorption, vapor condensation.
- Wave-pore size analysis: Electroacoustic /Conductivity.

From the porosity measurements, the following values can be calculated:

- Effective Porosity
- Bulk Volume Irreducible data (BVI) and Free-Fluid Index
- Pore Size Geometry
- Pore Size Distribution
- Melt Index Porosity
- Adsorption Pore Size Distribution
- Micro Pore Surface Area

For woven fabric, the fabric porosity depends on the number of interlacements of the weft and warp yarns. The relation between the air permeability and porosity is highly correlated [32–34]. Their values are important to be measured, depending on fabric application.

8.5.3.4 Fabric Strength

8.5.3.4.1 Uniaxial Strength of Fabric

Universal testing machine is used for the fabric strength determination. The speed of testing varies from 0.001 mm/min up to 1000 mm/min. The testing fabric samples are cut according to the standards (D4964 – 96(2012), D5034 – 09(2013), D5035 – 11(2015)) and gripped in two jaws, one is fixed on the frame while the other moves down at a predetermined speed. The fabric may be tested under constant rate of traverse or constant rate of loading. The data of the test is automatically recorded and analyzed. Figure 8.20 shows the load- extension curve of fabric tested under the uniaxial load.

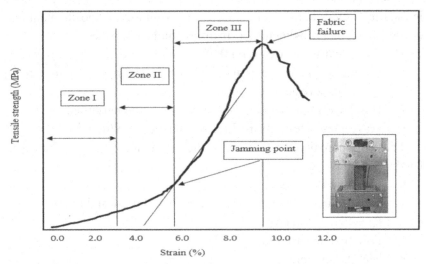

FIGURE 8.20   Load elongation curve of fabric.

The mechanical properties of the fabric, such as breaking load "N," breaking extension percentage, Initial modulus "N," and Work of rupture "N.m," can be obtained directly on the result sheet of Universal testing machine. Sample raveled strip dimension is 50 mm width × 150 mm in length, unraveled fabric width 25mm. Strip sample dimension is 25 mm in width × 150 mm in length. All tests must be performed under Standard atmosphere for conditioning relative humidity 65% rh and temperature 20°C. The fabric load extension curve, Figure 8.20, has three different behaviors, depending on the fabric extension under the load: in the first zone the fabric extended under low values of loads to overcome friction between the un-straight fibers in the weft and warp yarns in the direction of the load, in the second zone, the de-crimping zone, where the yarns in the direction of the load are trying to straighten and remove the crimp, till they reach the jamming condition, in the third zone the load will extend the yarns to its failure. This mechanism of fabric failure represents a threat to the failure mechanism of the composite, due to the assumption that the reinforcement is strained lower than that of the matrix, which is true only in the third zone. Fabrics with high strain in zones one and two will have a micro cracks at the low values of loading. Also, during the de-crimping zone, the crimp interchange mechanism will apply a compression stress into the composite during tensing it. The designer of the fabric for the laminate should take this into considerations and it is recommended to use low crimp fabric with stiff yarns. The mechanism of yarn failure in a fabric subjected to uniaxial tensile testing is strongly influenced by fabric geometry at the instant of failure and by yarn properties, both mechanical and topological [35].

To convert the fabric breaking load value in Newton "N_{fabric}" to MPa the following equation can be used:

$$\text{Fabric strength (MPa)} = 10^3 \, (\Phi.\rho_f/E.B)(N_{fabric}/tex_{yarn}) \qquad (16)$$

where: tex – yarn count "tex_y," Φ – yarn packing coefficient, ρ_f – fiber density "g/cm³," E – numbers of the ends/cm in the fabric, B – width of the tested fabric specimen cm.

Example: If the fabric strength 800 N, (E.B) 70, yarn count 25 *tex*, the fabric strength in MPa will be: 514.2 MPa.

The woven fabric strength is a function of the yarn strength, its surface morphology, coefficient of friction as well as the fabric structure. The strength of fabric stripe per yarn is in the range of 1.1 for ring spun yarns and to 1.5 for open end and air jet yarns.

8.5.3.4.2 Biaxial Strength of Fabric

In some applications, the fabrics and composites are under a biaxial loading, where the load P_w is applied in the warp direction and the load P_f in the weft direction, as illustrated in Figure 8.21.

The biaxial loading leads to different stress distribution on the various parts of the material, and the failure mechanism will be different than for the uniaxial loading. The testing specimen in the case of biaxial loading is cross shaped [36]. The stress distribution on the test specimen during loading is shown in Figure 8.21, from which it is clear that the strain distribution is varied along the area of the test specimen. For anisotropic properties of fabric, more strain was noticed at the edges.

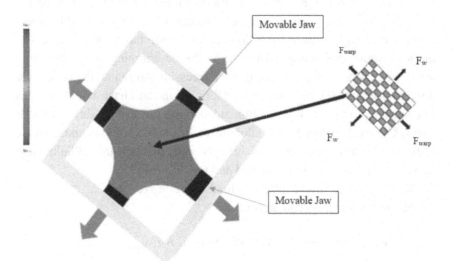

FIGURE 8.21   Strain map of biaxial loading.

8.5.3.5 Fabric Stiffness

The fabric stiffness is determined by several methods, of which Cantilever bending is the most common (ASTM D1388). Figure 8.22 shows the principle of this method where a stripe of fabric (fabric sample dimensions 25 mm × 150 mm) is slide on smooth horizontal surface. The fabric stripe will bend under its own weight when its outer edge touches the surface of indicator which is inclined 45° to the horizontal. The overhang length of the fabric is recorded and the fabric stiffness is calculated by the following equation:

$$C = l \, (\cos 0.5 \; \theta/8 \tan \theta)^{0.333} \tag{17}$$

where: l – the fabric length, θ – the bend angle of fabric.

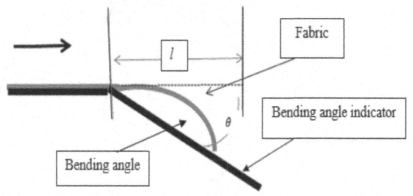

FIGURE 8.22   Principle of fabric stiffness measurement.

A new method of determining the fabric stiffness is based on the automatic bending of a certain fabric's length on device TH-7 [37] as illustrated in Figure 8.23. The force acting on one end of fabric while the other end is fixed, required to bend the fabric so it crosses the vertical line, is proportional to the fabric stiffness.

The bending stiffness of the fabric or composite is given by:

$$B = k \, (F_m/s) \tag{18}$$

where: B – bending stiffness Nm, k – a constant = 0.0334, F_m – bending force N, s – specimen width.

The device measures circular samples with the diameter of 5 cm. However, the device can be also used for measuring square or rectangular samples, and even for measuring samples of other shapes, such as triangular or trapezium. The fabric stiffness has impact on the delamination phenomena when the composite is subjected to the bending force as illustrated in Figure 8.24, hence the interfacial layers will undergo high shear force as the difference in the reinforcement and matrix stiffness increases and micro crakes may propagate, especially in the presence of voids.

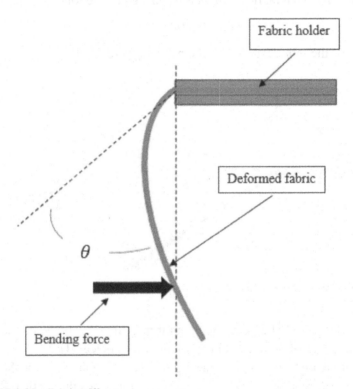

FIGURE 8.23 Fabric stiffness measurement.

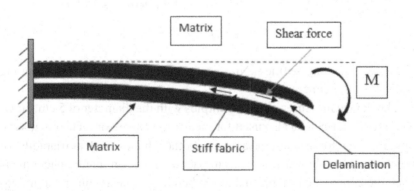

FIGURE 8.24 Matrix under bending moment.

8.5.3.6   Fabric Shear

The shear characteristic of a fabric determines the ability of a fabric to conform to a double-curvature surface which depends mainly on the in-plane shear behavior [38]. The shear properties of fabric associated with the biaxial state of stress in the fabric design: fabric designs with high number of yarn intersection show high value of shears resistance. Fabric shear properties measurements may be ether the direct method of measuring the shear force on the fabric and the shear angle or picture frame shear test. Picture frame shear test is carried out on the universal tensile testing machine by fixing the fabric in frame as shown in Figure 8.25a, which allows the sides of the fabric to move opposite to each other when tension load is applied on one end of the frame.

Forces act on the sample during the loading, Figure 8.25b, therefore causing in-plane fabric shear. The force and the angle of the shear are recorded to calculate the shear stress and shear modulus.

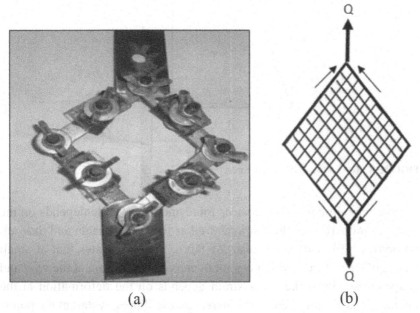

(a)   (b)

FIGURE 8.25   Picture frame shear test stand. (a) Shearing frame device (b) Forces acting on the sample in shear frame.

Figure 8.26 represents forces acting on the sample. Shear property was tested using a biaxial tensile fabric tester: the frame with the fabric fixed in two jaws of the tester in one direction. The relation between the frame pulling force F and the fabric shear property is given by the following equations:

$$\text{Shear stress } \tau = F/A \tag{19}$$

$$\text{Shear strain} = \Delta l/L \tag{20}$$

$$\text{Shear modulus} = (F/A)/tan\theta \tag{21}$$

where: F – shear force, A – area of the fabric cross section, τ – shear stress.

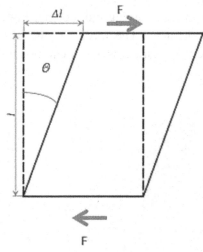

FIGURE 8.26   Fabric under shear force.

The relation between the shear force and shear angle depends on the physical properties of the fabric as well as the fabric design and the yarn properties [39]. The mechanism of fabric shear indicates that at small values of shear force, which cannot overcome the friction at the weft and warp intersections, the shear strain depends on the deformation at the intersections. As the shear force increases, the slippages start at the points of intersection overcoming the frictional forces, following by the elastic deformation till the complete fabric jamming. Figure 8.27 illustrates the relation between shear forces and shear angle for two different fabric designs.

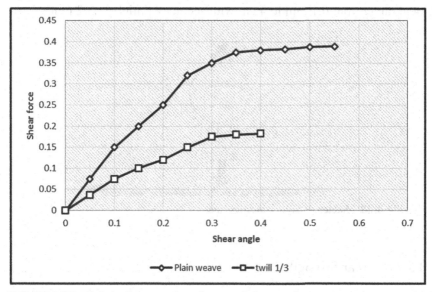

FIGURE 8.27   Relation between shear forces and shear angle.

8.5.3.7 Fabric Thickness

Fabric thickness is one of the basic physical properties of fabric speci-fications that influences the final composite properties. The fabric of the same weight per unit length can be produced having different fab-ric thickness, depending on several factors: yarn density, yarn packing density, fabric cover factor, fabric tightness and fabric design. There are several methods to determine the material thickness, such as the Gravimetric Methods, Stylus Method, Quartz Oscillator Method, and Optical Methods. The Stylus Method, Figure 8.28, is the most popu-lar for measuring the fabric thickness, according to ASTM D1777 – 96(2015).

Several other non-contact thickness gauges for measurement a textile material thickness depend on the electrical principles. Ultrasonic thickness gauge, dual – Lasers sensor thickness gauge, etc. are available. Ultrasonic

thickness gauge is used to measure thick fabric or composite on the basis
how long it takes for a sound pulse, caused by a small an ultrasonic trans-
ducer, to travel through a test piece and reflect back. Thickness measurement range from 0.1 mm to 4 cm with accuracy limits ±0.002 mm.

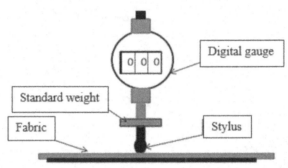

FIGURE 8.28   Sketch of Stylus thickness meter.

8.5.3.8   Tear Strength

Rectangular specimen 75 mm by 203 mm, slit at one end by the length of
75 mm, is placed into the universal testing machine – one side of the slit
end is clamped into the upper jaw and the other is clamped into the lower
jaw as shown in Figure 8.29a. The jaws move apart at a constant rate until
the fabric begins to tear. The tear speed is 50 mm/min. The highest peak
force needed to continue a tear reflects the strength of the yarns, fiber
bonds, or fiber interlocks. Trapezoidal sample, as shown in Figure 8.29b,
may be used for the determination of the tear strength.

   The load extension curve is illustrated in Figure 8.30, expressing that
the load increases till tearing of a yarn takes place followed by the increase
of extension till the jamming of the next yarn.

8.5.3.9   Puncture Strength

Testing puncture strength tests are used to determine the puncture of a
fabric. This is generally a test where a material is punched by a probe or
other type of sharp element until the fabric allow the probe to penetrate

through. The test is performed on the universal testing machine using special holder for the fabric (ASTM F1342/F1342M). Figure 8.31 represents the punching attachment on the universal testing machine and the punched sample. The loading of the specimen is CRL. The puncture test may be also carried out through impact by a falling weight, according to ASTM D5420. Figure 8.32 shows a setup for fabric falling weight tester. Dropped weight impact for penetration resistance tests was performed to simulate a knife or blunt spike stab impact on the fabric on a specially designed apparatus, as shown in Figure 8.32. Dropped weight-impact stab tests are performed with the primary objective to identify the maximum energy needed to puncture the fabrics. The punching speed will be the speed of the spike at the instance of impacting the sample surface.

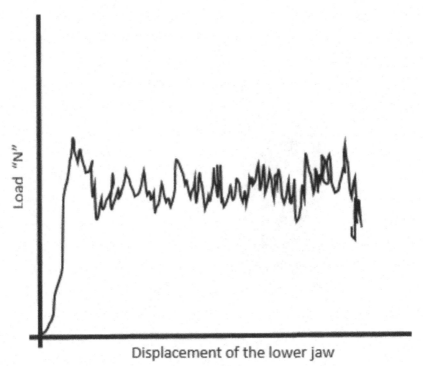

FIGURE 8.30 Fabric tearing test in weft direction.

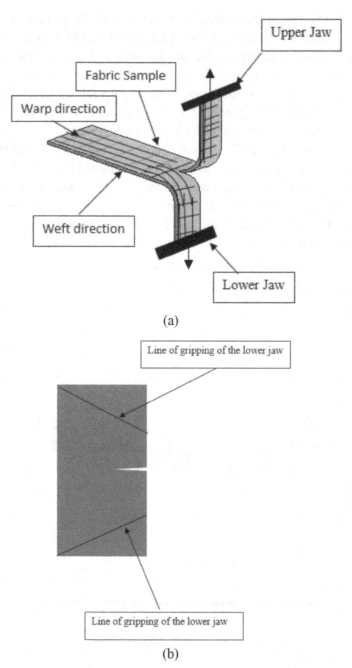

FIGURE 8.29 Tear test samples.

FIGURE 8.31 Punching attachment on universal testing machine.

The falling weight test is used to fulfill the compliance of NIJ Standard-0115.00 requiring protection against low energy, medium or high energy threats, so the fabric should withstand strike energy with standard spike or knife of 24 J, 33J and 43J, respectively [40].

Analyses of the punching mechanism indicated the following possible scenarios:

Scenario A, impactor's edge punches the fabric between the threads, hence sufficient space allows it to pass, as in the case of fabric with low density in weft and warp. The yarns will move in different directions permitting impactor edge to pass through without fiber or yarn damage.

Scenario B, impactor's edge punches the fabric between the threads without cutting matrix, pushing the yarns aside without cutting it.

Scenario C, for tight fabric near the jamming condition, the yarns start to be cut by impactor's edge. If the strike energy is higher than fabric resisting energy, the impactor will pass completely though the fabric. Impactor's edge perforates the fabric through cutting the yarns in contact area with its blade edge. In this case, part of the energy is absorbed by deformation of the fabric; the other part is absorbed through the cutting of number of yarns to allow impactor's edge to pass through.

Scenario D, if impactor's edge cannot perforate the fabric, as in the case of blunt impactor, the fabric will be deformed causing the yarns to distort and travel with the same speed of the falling impactor.

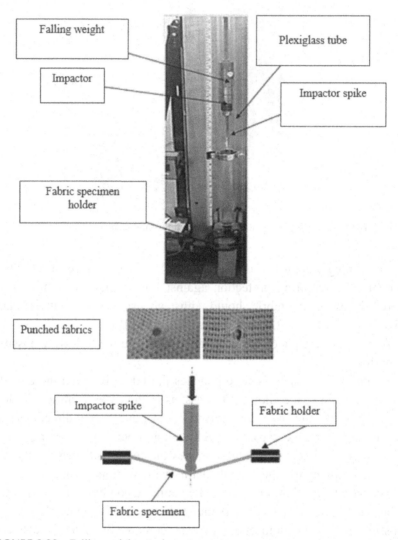

FIGURE 8.32 Falling weight attachment.

This will form a cone transferring the strain to all other yarns, either in the same direction or in different directions. The failure will take place only if the strain in the yarns reaches the breaking strain.

In the case of the natural fiber polymer composite, the Scenarios C and D are the most probable.

In Scenario C, the matrix starts to be cut by impactor's edge, if the strike energy is higher than matrix resisting energy, then impactor will pass completely though the matrix layer and impactor's edge perforates the fabric through cutting the yarns in contact area with its blade's edge.

In Scenario D, the impact may cause the delamination between the laminates or fibers or yarns, if the binding energy of the composite components is less than the impact energy.

The straight way to evaluate the impact performance of a fabric is to calculate its energy absorption. Hence, the kinetic energy of the impactor before impact ($E_{before\ impact}$) will be equal to the sum of the kinetic energy absorbed ($E_{absorbed}$) by the fabric and kinetic energy of the impactor after impact ($E_{after\ impact}$). Therefore:

$$E_{absorbed} = E_{before\ impact} - E_{after\ impact} \tag{22}$$

The energy lost during impact $E_{absorbed}$ is given by:

$$E_{absorbed} = 0.5\ M\ (V_1{}^2 - V_2{}^2) \tag{23}$$

where: $E_{absorbed}$ is the kinetic energy absorbed by the fabric, V_1 is the velocity of the impactor before impact, V_2 is the velocity of the impactor after impact, and M is the mass of the impactor.

Kinetic energy absorbed by the fabric $E_{absorbed}$ is defined by the following six different components:
1. ES: Energy to shear the yarns.
2. ED: Energy to deform all other yarns.
3. ET: Energy to tensile failure of directly impacted yarns.
4. EF: Energy to overcome friction between fabric layers.
5. EJ: Energy to overcome friction between blade and yarns.
6. EM: Energy to move the fabric during impact.

Therefore, the total absorbed energy by the fabric is expressed as the sum of all the above energies. In the case of the composite of fabric laminate, yarns are fixed in position due to the presence of matrix. This will lead to the energy increase required to punch the composite due to the higher values of EM, EJ, E and ET. The percentage of increase of the total energy depends on the polymer properties and the value of the interfacial bonding between the yarns and the matrix.

8.5.3.10   Bursting Strength

Pressure at which a fabric or sheet plastic composite will burst is used as a measure of resistance to rupture, burst strength depends largely on the tensile strength and extensibility of the material. There are two method for the measuring the fabric bursting strength: the measurement of the bursting strength by means of a ball burst mechanism or Diaphragm. Bursting ball bursts the fabric which is fixed in the lower jaw of the universal testing machine while a ball attachment is fixed on the upper jaw as shown in Figure 8.33. The bursting of the ball in the fabric will continue till the complete penetration takes place (ASTM D6797). The applied load on the fabric may be CRT or CRE, which gives different results. The burst strength is expressed in MPa and is determined by the formula, for ball diameter 2.54 cm:

$$\text{Bursting strength "MPa"} = 0.001974\ F_{bursting} \tag{24}$$

where: $F_{bursting}$ is bursting force "N."

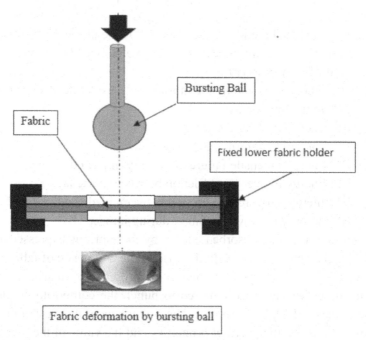

FIGURE 8.33   Sketch of ball bursting strength tester.

In diaphragm bursting strength tester, the sample is fixed on holder over the rubber diaphragms, shown in Figure 8.34. Under the increase of the hydraulic pressure, the rubber diaphragm strains the sample till bursting of the fabric occurs (ASTM D3786/D3786M). It can be used for knitted, nonwoven and woven fabrics. The bursting pressure of each specimen can be calculated by subtracting the pressure required to inflate the diaphragm from the total pressure required to rupture the specimen.

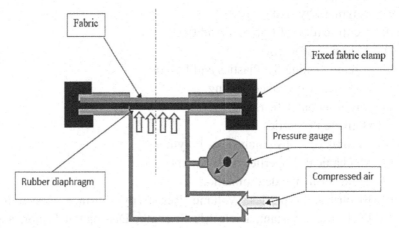

FIGURE 8.34   Sketch of diaphragm bursting strength tester.

8.6   POLYMER TESTING METHODS

The polymers with the different properties, depending on the polymer length of the polymer chain and chain structure, are used to form the matrix. There are thermoset and thermoplastic polymers. When the thermoplastic polymer is heated, the de-bonding between chains occurs and the viscosity of the polymer increases, on the contrary to thermoset polymers – on heating of the polymer no movement between the molecules.

8.6.1   PROPERTIES OF POLYMER

The following polymer properties should be known to the designer to assist in the selection of the suitable matrix for a certain application:

1. Intrinsic viscosity measurement
2. Density measurement
3. Chemical family
4. Tensile properties
5. Flexural strength
6. Thermal – mechanical strength
7. Compression
8. Creep properties
9. Polymer physical properties
10. Identification of Polymer Additives
11. Adhesive properties
12. Ageing Testing for Plastics and Polymers
13. Chemical Resistance Testing
14. Environmental Testing
15. Ballistic Properties
16. Chromatography Analysis of Polymers
17. Mechanical Properties of Polymers
18. Mold shrinkage determination
19. Electrical Properties: Volume Resistivity, Surface Resistivity, Dielectric Constant, Dielectric Strength, Dissipation Factor, and Arc Resistance.

8.6.2   TESTING METHODS FOR POLYMER MATRIX COMPOSITE MATERIALS

Depending on the application of the polymer and the composite end use, the test should be chosen. Table 8.7 gives some of the basic tests that provide a required knowledge to a designer of composite material.

8.7   COMPOSITE MATERIALS TESTING METHODS

8.7.1   INTRODUCTION

Composite testing plays a decisive role across the composites supply chain and product life cycle. According to the expected forces which probably applied during the service of the composite, it will be tested before accepted for an application. The attachment used to apply the load for testing will follow the recommendations of the standards, such as ASTM, BS, and ISO. The type of test is trying to imitate the loading condition during the use of a composite part. The success of the use of composite is to know exactly fracture mechanics of the designed part, thus a suitable flexure is designed to imitate the loading condition mode. For instance, assuming the following fraction modes in the case of composite delamination, as shown in Figure 8.36, to determine propagation characteristics of existing cracks.

Mode I: Opening or extension;

Mode II: Shear;

Mode III: Tearing or twist.

TABLE 8.7 Polymer Testing Methods

Test	Method	ASTM	Ref.
Intrinsic viscosity measurement	1. Solvent-free method based on the extrusion of the polymer through a die using a dedicated instrument 2. Solvent-based techniques 3. Free blowing method	D4603	
Density Measurement	Five samples are prepared by cutting approximately 6.5 mm from a preform, then dropped into a density gradient column and allowed to settle for approximately 15 minutes. The height of the samples is then carefully measured. Standard density balls are used to calibrate the density of the column. The density of polymers varied between 0.57–0.68 g/cm³ for Polypropylene Uniboard Ultrastiff to 3.9 g/cm³ for PTFE Tetron B.	D792	
Tensile properties	The force required to break a polymer composite specimen and the extent to which the specimen the test speed can be determined by the material specification or time to failure (30 sec to 5 minutes) measure. Sample size: 25 mm wide and 115 mm long, (10 mm) thickness. Testing speed 0.125–500 cm/min	D3039 D5083 D638 D638	

TABLE 8.7   (Continued)

Test	Method	ASTM	Ref.
Chemical family	Spectroscopy FTIR, Fourier Transform Infrared. Infrared radiation is passed through a sample. Some of the infrared radiation is absorbed by the sample and some of it is passed through (transmitted). The resulting spectrum represents the molecular absorption and transmission, creating a molecular fingerprint of the sample. For each polymer there is standard FTIR spectrum, as shown in Figure 8.35. By FTIR spectrum of the organic materials bulk and small particle materials can be analyzed.	E1252 E168	[41, 42]
Flexural strength	Three-point bending test with center loading on a simply supported beam or four-point bending test with two loading points. Standard universal testing machine.	D7264	
Thermal properties	The thermal properties of the polymer include the effect of the heat on their physical and mechanical properties such as viscosity, plasticity, creep, strength compressibility, fatigue, flexure strength, which depends on polymer thermal conductivity, melting point, specific heat capacity.		
Compression	Compression test determines the mechanical properties of unreinforced and reinforced rigid plastics, including high-modulus composites, when loaded in compression at relatively low uniform rates of straining or loading. The application of compression load on sample is found to use several shapes of fixture, generally the specimen is tested under compression or combined load using one of the following methods: Shear loading method, End loading method or Combined loading method.	D6641 D3410 D695	[43, 44]
Creep	This test method measures the creep rupture time of FRP bars under a given set of controlled environmental conditions and force. The ranking of the different polymers indicates that Polyester and nylon show high resistance to creep on the contrary to Polyethylene.	D7337	

TABLE 8.7 (Continued)

Test	Method	ASTM	Ref.
Glass Transition Temperature, Tag	Temperatures indicate change in the matrix material from the glassy to the rubbery state during heating. A change in matrix stiffness of two or three orders of magnitude occurs during the glass transition. Dynamic Mechanical Analysis (DMA) is commonly used. Tag value is usually used to indicate the upper use temperature of composite materials.	D7028	
Thermal expansion properties	The coefficient of thermal expansion (CTE) values are of considerable interest to design engineers. It defines the rate at which a material expands as a function of temperature. Method for the measuring of CTE includes mechanical dilatometry, optical imaging and interference systems, x-ray diffraction methods and electrical pulse heating techniques. Polyethylene has the highest value of coefficient of thermal expansion. While PAI Polyamide-imide has the smallest value.	D696	[45]
Deflection temperature under load	The temperature of the loaded beam is raised until a certain amount of deflection is observed. The temperature when that deflection is reached is called the DTUL. The DTUL is sometimes referred to as the Heat Distortion Temperature or HDT.	D648–01	
Moisture absorption	As moisture is absorbed into a polymer composite it has a plasticizing effect on the material and a degradation in the strength of the composite. The moisture absorption of polymer varied "between" 0.01% to 1.9%.	D5229	[59]
Chemical resistance	Chemical resistance of the polymer is highest for PE, Polyethylene, moderate for Polyester and low for PC Polycarbonate.		
Flammability	Flammability Oxygen Index of polymers varied between 58 and 15. Polyester and nylon have moderate values.	E2058–13a	
Impact resistance	Polypropylene has the highest value of impact resistance while Polyester is the worst.	D7136	

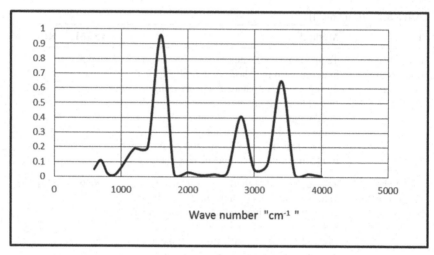

FIGURE 8.35   FTIR spectrum.

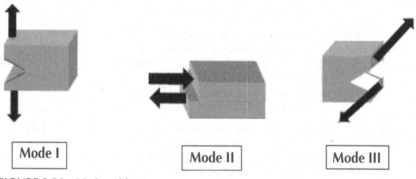

FIGURE 8.36   Modes of fraction.

8.7.2   *TESTING METHODS FOR COMPOSITE MATERIALS*

The survey of testing methods and the suitable attachment designed
to test preform in the different modes of composite failure is shown in
Table 8.8.

TABLE 8.8 Attachments for Composite Testing

Mode	Procedure	ASTM	Ref.
Mode I			[42,43]

FIGURE 8.37 The specimen will be split at its end andt attached to the testing machine as illustrated in Figure 8.35 to measure inter laminar fracture.

Mode II			[42]

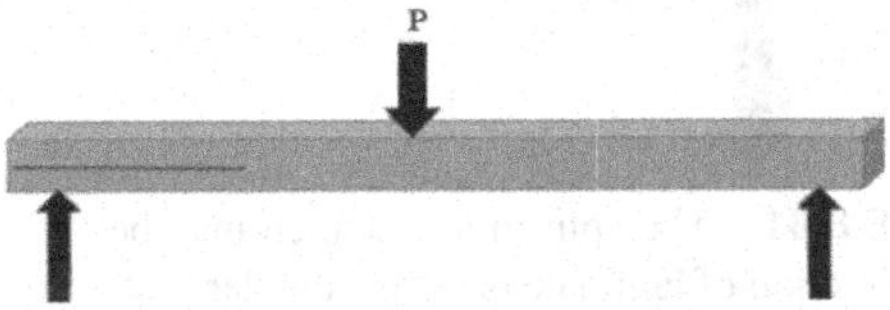

FIGURE 8.38 The specimen will be split at its end and attached to the testing machine as illustrated in Figure 8.36 to perform 3 point bending test. Both shear and tensile stresses are acting in this test.

Mode III			[46]

FIGURE 8.39 The split at the edge specimen is subjected to moments in all three axes.

Mixed mode I+II			[45]

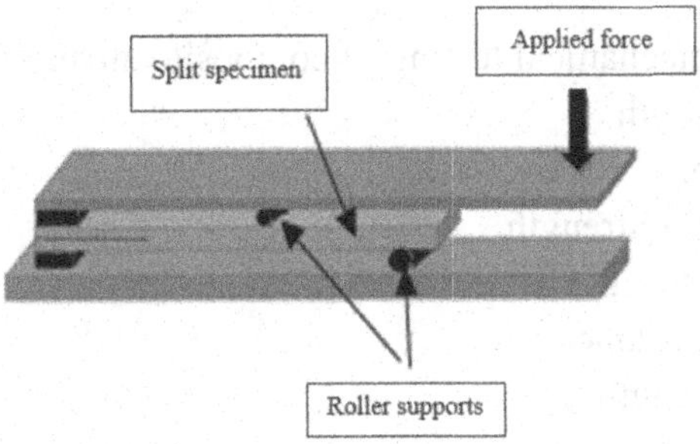

FIGURE 8.40 In this test the split specimen is subjected to the combination of Mode I and Mode II.

TABLE 8.8 (Continued)

Mode	Procedure	ASTM	Ref.
Intraluminal crack resistance on tension	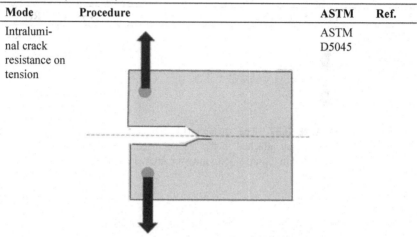	ASTM D5045	

FIGURE 8.41 The split in a specimen may be in the direction of laminate or perpendicular to it.

| Intraluminal crack resistance on bending | 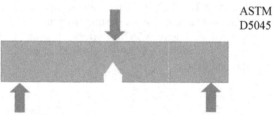 | ASTM D5045 | |

FIGURE 8.42 The notch in specimen may be in the direction of laminate or perpendicular to it. Tested on three point bending test, as shown in Figure 8.4.

Generally, the mechanical testing of composite material includes:

1. Tensile Strength
2. Compression
3. Flexure/Bend Strength
4. Puncture Strength
   - Tear Resistance
   - Peel Strength
   - Shear Strength
   - Delamination Strength
   - Bond Strength

- Adhesion Strength
- Creep and Stress Relaxation
- Crush Resistance
- Impact Strength
- Torsion

Table 8.9 gives some testing methods usually carried out on composite to characterize composite's properties which depends on the end use.

TABLE 8.9   Composite Testing

Test	Method	ASTM	Ref.
Moisture absorption	The specimens are dried in an oven for a specified time and temperature and then placed in a desiccator to cool. Immediately upon cooling, the specimens are weighed. The material is then emerged in water at agreed upon conditions, often 23°C for 24 hours or until equilibrium. Specimens are removed, patted dry with a lint free cloth, and weighed.	D570	
Impact	1. Out-of-plane impact force is one of the major concerns of many structures made of advanced composite laminates. Drop-weight impact testing is used for the determination of the damage response parameters and can include dent depth, damage dimensions, and through-thickness locations. Impactor geometry has a blunt, hemispherical striker tip.	D7136/ D7136M D3763 D 5628 D256 D1822	

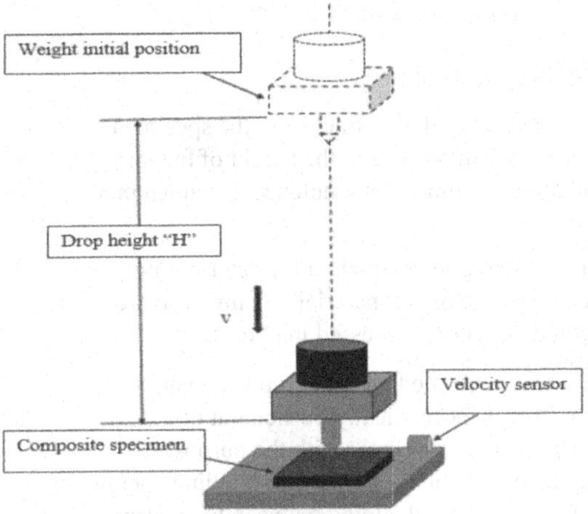

FIGURE 8.43   Free weight drop tester.

TABLE 8.9 (Continued)

Test	Method	ASTM	Ref.
	In drop weight impact testing, a mass is raised to a known height H and released, impacting the specimen.		

2. A notched specimen of dimensions $63.5 \times 12.7 \times 3.2 \ mm^3$ (with the notch facing away from the point of contact) is placed into an instrument with a pendulum of a known weight. The pendulum is raised to a known height and allowed to fall. As the pendulum swings, it impacts and breaks the specimen, rising to a measured height.

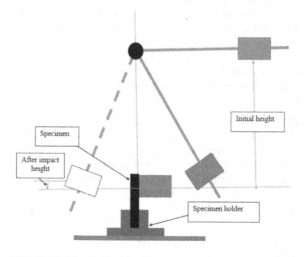

FIGURE 8.44 Ballistic tester.

Amount of energy lost due to fracturing the specimen is proportional to the difference in the height of the mass before and after the impact. Not notched specimen may be used.

3. Izod impact strength test method for determining the impact resistance of the materials is similar to the above method. Result is expressed in (J/cm).

| Fiber push-out test | Mechanical test performed on the composite materials where a fiber is mechanically pushed out of the material. This test is carried out with the purpose of measuring the matrix/fiber interface de-bonding energy and the effect of frictional sliding between the matrix and the fiber. | | [50] |

TABLE 8.9 (Continued)

Test	Method	ASTM	Ref.
Fiber pull-out	Fiber pull-out is one of the failure mechanisms in fiber-reinforced composite material. Other forms of failure include delamination, intraluminal matrix cracking, longitudinal matrix splitting, fiber/matrix de-bonding, and fiber fracture. The cause of fiber pull-out and delamination is weak bonding. $$W_d = \pi \, d^2 \, \sigma_f^{\,2} \, l_d / 24 \, E_f$$ where: d is fiber diameter, σ_f is failure strength of the fiber, l_d is the length of the de-bonded zone, and E_f is fiber modulus.		[51]
Tensile testing	Tensile testing is a fundamental materials science test in which a sample is subjected to a controlled tension until failure. The results from the test are commonly used to select a material for an application, and to predict how a material will react under other types of forces. Specimen dimensions: 25 mm wide and at least 250 mm long. Thickness can be between 2 mm and 14 mm. Test speed is 2–5 mm/min.	D882 D5083 D3039/ D3039M	
Adhesive strength	Representative of actual joint to be used. Iosipescu Shear test or bonded joint test can be used. Thick adhered specimen, as shown in Figure 8.46, is tested on tension to measure the bonded joint characteristics.	D 5379 D 5656	[51]

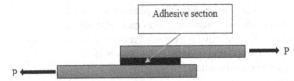

FIGURE 8.46 Bonded joint test.

In Figure 8.46, P – load per unit width of joint; l – specimen width; t – sheet thickness; σ – P/l.

$$\tau = \sigma \, (t/l)$$

where τ – shear stress in the adhesive.

TABLE 8.9 (Continued)

Test	Method	ASTM	Ref.
Compressibility	The objective of this tests is to determine the compressive properties, such as: • Ultimate compressive strength • Ultimate compressive strain • Compressive chord modulus of elasticity • Compressive Poisson's Ratio Specimen should have a uniform rectangular cross section, 140 mm to155 mm long, width can be 12 mm or 25 mm. various thickness can be tested by applying the compressive force into the specimen.	D3410	
Density measurement	Measure the density of solid materials.	D1505–68, D 792	
Compression	In compression test is applied to measure compressive strength and stiffness of polymer matrix composite materials using a combined loading compression test fixture. The other test procedure designed to measure the compression residual strength properties of multidirectional polymer matrix composite laminated plates. Exposure to damage from concentrated out-of-plane forces. Damage resistance and damage tolerance properties that are derived from this test are useful for composite material design. Several loading fixture were designed to test the samples on compression. This test procedure introduces the compressive force into the specimen through combined shear end loading. The specimen geometries also differ markedly from method to method. Strip of composite material having a constant rectangular cross section. The specimen is constrained from buckling by sandwiching it between lateral supports that are lightly bolted together.	D 6641/D 6641M-01 D 695–96 D7137 D 3410	
Shear	Several methods are used for measuring the shear properties of the composites. I. In-plane shear 1. Iosipescu Shear 2. V-Notched Rail Shear 3. ±45 Tension Shear	D5379 D7078 D3518 D3846 D5379	[48] [49]

TABLE 8.9 (Continued)

Test	Method	ASTM	Ref.
	II. Out-of-plane shear		
	1. Short Beam Shear		
	2. Iosipescu Shear		
	Figure 8.47 shows the principles of Iosipescu Shear tester.		

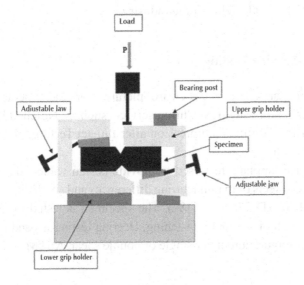

FIGURE 8.47 Iosipescu Shear Tester.

| Fire calorimetry | The cone calorimeter is used to test fire performance at radiant heat fluxes of 25, 50, 75, and 100kW/m². MIL-STD-2031 fire response parameters include the peak and average HRRs (kW/m²) and the time to ignition (Tig) (seconds) as well as the type and amount of combustion gases produced at a 25kW/m² incident heat flux. | E 1354 | [52] |
| Flammability | The flammability test is used to determine the relative rate of burning of self-supporting plastics. The following parameters are measured: 1. Time until the flame extinguishes itself. 2. Distance the burn propagates. 3. Linear burning rate in mm per minute. | D635 | |

8.7.3 OTHER COMPOSITE MATERIAL TEST METHODS

Several testing procedures may be carried out in order to give the designer an idea about the performance of the composite when loaded as a part of the complicated structure and the deterioration in the composite parts during the service life. The testing types and procedures of composite material have been on the increase in the last decade due to the diverse types, shapes and composition of the composite materials to satisfy the vast applications in all fields of the industry.

8.7.3.1 In-Service Testing

These tests are carried out to get clearer picture about the expected behavior of the final composite parts during service, such as: Notched Laminate Testing, Bearing Testing, Compression after Impact Testing, and Fracture Mechanics Testing.

The notched laminate test used to provide design values mechanically fastened joints effects of manufacturing differences and small damage areas, ASTM standards (D 5766, D 6484). The specimen is tested in tension or compression with or without a fastening. Bearing testing is used to specify bolted joint configuration under single or double shear, ASTM D 5961.

8.7.3.2 Nondestructive Tests (NDT)

The nondestructive tests are very efficient method for the detection of the Inhomogeneity within the materials, the presence of fiber breakage, the presence of resin micro-cracking, percentage of voids, the change of porosity, delamination and other manufacturing defects. Also, the composites during their service life are subjected to fatigue, impact, creep loading and change of environmental conditions, temperature, and humidity. These loads can cause great damage to the composite (cracks, delamination, and fiber matrix de-bonding). Nondestructive test is required to inspect the composite in-service. There is a long list of NDT methods and sub techniques that are applicable for composite testing [47]. Table 8.10 gives principles of some Nondestructive tests.

TABLE 8.10 Principles of Some Nondestructive Tests

Test	Method	ASTM	Ref.
Ultrasonic	There are several apparatus build on the principle of using sound inspection of composite using a low frequency sound wave. Ultrasonic testing commonly used as non-distractive inspection method for composite materials. Short pulses of ultra sound are passed through the composite and detected after having interrogated the structure and analyzed.	E2533 E2580 – 12	[49]
Method of application Ultrasonic test			
	Pulse- echo: a pulse of ultrasonic energy is transmitted into the specimen in perpendicular to its surface. The amplitude of the echo will give an idea of the size of the defect. The specimen is usually immersed in water during testing.		[53, 54]
	Back scatter: the transducer is inclined at an acute angle to normal to test specimen surface. During measurement, the transducer rotates around the same angle. The maximum back signal when the transducer is perpendicular to the direction of the fibers in the composite.		[55]
	Through – transmission: in this case two transducers are used, one emitter and the other receiver. Water jet at each transducer is acting as wave guide.		[56]
	Ultrasonic spectroscopy: the pulse arrived to the receiver transducer is analyzed harmonically, thus detect the characteristic of the composite structure.		[56]
	Acoustography: the receiver probe is focused into system of lenses after passing through the specimen.		[56]
X-Ray Inspection	The X-ray inspection technique is used since long time to detect density changes and well suited for bonded interfaces and can locate delamination, presence of voids, change of porosity, inclusions.		[48]
Thermography	The principle of this method depends on heat transfer through the composite which is recorded using infrared video camera to allocate the thermal changes in the composite cross section.		[57]

TABLE 8.10   (Continued)

Test	Method	ASTM	Ref.
Shearography	Sheargraphy is a laser based optical measuring technique that provides fast and accurate indications about internal material discontinuities or irregularities in non-homogenous materials.		[58, 59]
	Using laser light, a shearing interferometer is able to detect extremely small (sub-micrometer) changes in surface out-of-plane deformation. When a test object is subjected to an appropriate load, a proportional strain is induced on the test surface. If underlying discontinuities are present, the surface will deform unevenly at these locations. This is then interpreted through the shearing interferometer as a change in the phase of the laser light.		

All the NDT methods are used in product and process design optimization, on line process control, after manufacturing inspection as well as in-service inspection.

KEYWORDS

- **composites**
- **fabric**
- **fibers**
- **polymers**
- **standards for testing**
- **testing methods**
- **yarns**

REFERENCES

1. SIROLAN™ LASERSCAN, A New Technology for a New Millennium. [Online] file:///C:/Users/magdy/Downloads/Laserscan%20Technology%20Brochure%20(2). pdf (accessed May 12, 2016).

2. BSC Electronics. [Online] http://www.ofda.com/Natural_fibers/Ofda100.html (accessed May 12, 2016).

3. Qi K.; Lupton, C. J.; Pfeiffer, F. A.; Minikhiem, D. L. Evaluation of the optical fiber diameter analyzer (OFDA) for measuring fiber diameter parameters of sheep and goats. J. Anim. Sci. 1994, 72(7), 1675–1679.

4. ASTM D 2130 – 90, Standard Test Method for Diameter of Wool and Other Animal Fibers by Micro projection.

5. Moore, E. M.; Shambaugh, R. L.; Papavassiliou, D. V. Ensemble Laser Diffraction for Online Measurement of Fiber Diameter Distribution during the Melt Blowing Process. INJ. [Online] 2004. http://www.jeffjournal.org/INJ/inj04_2/p42–47-moore. pdf (accessed Dec.4, 2015).

6. Zeyun, C.; Rongwua, W.; Xianmiao, Z.; Baopu, X. Y. Study on Measuring Microfiber Diameter in Melt-blown WebBased on Image Analysis. Procedia Eng. 2011, 15, 3516–3520.

7. Oladele, I. O.; Omotoyinbo, J. A.; Adewara, J. O. T. Investigating the Effect of Chemical Treatment on the Constituents and Tensile Properties of Sisal Fiber. J. Miner. Mater. Charact. Eng. 2010, 9(6), 569–582.

8. Madell, J. F.; Chen, J. H., Mc Garry, F. J. A micro-debonding test for in situ assessment of fiber/ matrix bond strength in composite materials. International Journal of Adhesion and Adhesives 1980, 1(1), 40–44.

9. Nishikawa, M.; Okabeb, T.; Hemmia, K.; Takedac, N. Micromechanical modeling of the micro bond test to quantify the interfacial properties of fiber-reinforced composites. Int. J. Solid. Struct. 2008, 45(14–15), 4098–4113.

10. Sockalingam, S. Fiber-Matrix Interface Characterization through the Micro bond Test. Int'l. J. Aeronautical & Space Sci. 2012, 13(3), 282–295.

11. Loan, T. T. Investigation on jute fibers and their composites based on polypropylene and epoxy matrices. PhD Theses, Der Fakultät Maschinenwesen der Technischen Universität Dresden, 2006.

12. Yang, Li.; Thomason, J. L. Interface strength in glass fiber-polypropylene measured using the fiber pull-out and micro bond methods. Composites Part A 2010, 41(9), 1077–1083.

13. Herrera-Franco, P. J.; Drzal, L. T. Comparison of methods for the measurement of fiber/matrix adhesion in composites. Composites 1992, 23(1), 2–27.

14. Sain, M.; Suhara, P.; Law, S.; Bouilloux, A. Interface Modification and Mechanical Properties of Natural Fiber–Polyolefin Composite Products. J. Reinf. Plast. Compos. 2005, 24(2), 121–130.

15. Tresanchez, M.; Pallejà, T.; Teixidó, M.; Palacín, J. Measuring yarn diameter using inexpensive optical sensors, Proceedings Eurosensors XXIV, 2010, Linz, Austria.

16. Basua, A.; Doraiswamya, I.; Gotipamula, R. L. Measurement of Yarn Diameter and Twist by Image Analysis. JT I. 2009, 94(1–2).

17. Li, Z.; Pan, R.; Zhang, J.; Li, B.; Weidong Gao, W.; Wei Bao, W. Measuring the unevenness of yarn apparent diameter from yarn sequence images. Measurement Sci. Technol. 2016, 27(1), 1–10.

18. Eldessouki, M.; Ibrahim, S.; Militky, J. A dynamic and robust image processing based method for measuring the yarn diameter and its variation. Text. Res. J. 2014, 84(18), 1948–1960.

19. Ibrahim, S.; Militky, J.; Kremenakova, D.; Mishra, D. Characterization of yarn diameter measured on different. Proceedings RMUTP International Conference: Textiles & Fashion 2012, Bangkok Thailand, 2012.
20. Narkhedkar, R. N.; Kane, C. D. Study of yarn cross section shape and its diameter, Indian J. T. 2012.
21. Tsai, I. S.; Chu, W. C. The Measurement of Yarn Diameter and the Effect of Shape Error Factor (SEF) on the Measurement of Yarn Evenness. JT I. 1996, 87(3), 496–508.
22. Militky, J.; Sayed Ibrahim, S.; Kremenakova, D.; Mishra, R.; Muckova, E.; Mvubu, M. comparative study of yarn diameter measurements. 19 Strutex Conf. Liberec, Czech Republic 2009.
23. Uster Laboratory system application report. [Online] https://www.uster.com/fileadmin/customer/Knowledge/Textile_Know_How/Yarn_testing/U_LabSystems_Measurement_of_slub_01.pdf (accessed May15, 2015).
24. Keissokki Kogo Co. Ltd., Japan. Keisokki Laserspot Hairiness Diameter Tester and KET-80. Indian Tex. J. 2012, 122(9), 110.
25. USTER Evenness Tester 5 https://www.uster.com/fileadmin/customer/Instruments/Brosch%C3%BCren/en_USTER_TESTER_5_web_brochure.pdf
26. Huh, Y.; Suh, M. W. Measuring thickness variations in fiber bundles with flaying laser spot scanning method. Text. Res. J. 2003, 73, 767–773.
27. Voborova, J. Neckar, B. Ibrahim, S., Garg, A. Yarn properties measurements: an optical approach. The 5th international Engineering Conference of Mansoura University, Egypt, 2003.
28. Vlasblom, M. P.; Bosman, R. L. M. Predicting the Creep Lifetime of HMPE Mooring Rope Applications. Proceedings of OCEANS 2006, IEEE conf., Boston, Ma, USA, 2006.
29. Lokendra Pal, L.; Joyce, M. K.; Paul Daniel Fleming, P. D. A simple method for calculation of the permeability coefficient of porous media, TAPPI J. 2006, 5(9), 10–16.
30. Nazarboland, M. A.; Chen, X.; Hearle, J. W. S.; Lyndon, R.; Moss, M. Effect of different particle shapes on the modeling of woven fabric filtration, J. Inf. Comput. Sci. 2007, 2(2), 111–118.
31. Jena, A.; Gupta, K., Characterization of pore structure of filtration media. [Online] https://www.academia.edu/8460896/characterization_of_pore_structure_of_filtration_media (accessed April 15, 2015).
32. Abd El-Hakam, M. A Study of Some Parameters Affecting the Porosity of the Textile Structures. MSc. Theses, AU, 2012.
33. Epps H.; Leonas K. L. The Relationship between Porosity and Air Permeability of Woven Textile Fabrics, JTE 1997, 25(1), 108–133.
34. Elnashar, E. A. Volume porosity and permeability in double-layer woven fabrics. AUTEX Res. J. 2005, 5(4), 207–218.
35. Seo, M. H.; Realff, M. L.; Pan, N.; Boyce, M.; Schwartz, P.; Backer, S. Mechanical Properties of Fabric Woven from Yarns Produced by Different Spinning. Text. Res. J. 1993, 63, 123–134.
36. Galliot, C.; Luchsinger, R. H. Biaxial testing of architectural membranes and foils, TensiNet Symposium SOFIA 2010 – Tensile Architecture, Connecting Past and Future. Sofia, Bulgaria, 2010 [Online] http://w3.uacg.bg/ftp/tensinet2010/Photos_and_

Presentations/Presentations/Biaxial%20testing%20of%20architectural%20membranes%20and%20foils.pdf (accessed May15, 2015).

37. Fridrichová L. A new method of measuring the bending rigidity of fabrics and its application to the determination of their anisotropy. Text. Res. J. 2013, 83(9) 883–892.

38. Peil, K. L.; Barbero, P. J.; Sosa, E. M. Experimental evaluation of shear strength of woven webbings. SAMPE 2012 Conference and Exhibition, Baltimore, USA, 2012.

39. Sun, H.; Pan, N. Shear deformation analysis for woven fabrics. Compos. Struct. 2005, Vol. 67, Issue 3, 317–322.

40. El Messiry, M.; Eltahan, E. Stab resistance of triaxial woven fabrics for soft body armor, J. Ind. Text. 2016, 45(5), 1062–1082.

41. Yang, C. Chemical Analysis of Polymeric Materials Using Infrared Spectroscopy. [Online] http://www.t-pot.eu/docs/Yang_2011–11–2%20U%20Zagreb.pdf (accessed June 15, 2015).

42. Collection of IR spectra for some common polymers. http://www.doitpoms.ac.uk/tlplib/artifact/flash/infrared.swf (Accessed June 6, 2015)

43. Adams, D. Current compression test methods. Compos. World [Online] 2005, http://www.compositesworld.com/articles/current-compression-test-methods (accessed June 2, 2015).

44. Intertek. Composite Materials Test Methods. [Online] http://www.intertek.com/polymers/composites/test-methods/ (accessed June 2, 2015).

45. Ran, Z.; Yan, Y.; Li, J.; Qi, Z.; Yang, L. Determination of thermal expansion coefficients for unidirectional fiber-reinforced composites. Chinese J. Aeronautics 2014, 5 1180–1187.

46. Choi, N.S; Kinloch, A. J.; Williams, J. G. Delamination fracture of multidirectional carbon-fiber/epoxy composites under Mode I, Mode II and Mixed-Mode I/II loading. J. Compos. Mater.1999, 33, 73–100.

47. Prasad, M. S.; Venkatesha, S.; Jayaraju, T. Experimental Methods of Determining Fracture Toughness of Fiber Reinforced Polymer Composites under Various Loading Conditions. J.Minerals Materials Character. Eng. 2011, 10(13), 1263–1275.

48. Djordjevic, B. B. nondestructive test technology for the composites. The 10th International Conference of the Slovenian Society for Non-Destructive Testing, Application of Contemporary Non-Destructive Testing in Engineering. Ljubljana, Slovenia, 2009.

49. Djordjevic, B. B. Ultrasonic characterization of advanced composite materials. The 10th International Conference of the Slovenian Society for Non-Destructive Testing, Application of Contemporary Non-Destructive Testing in Engineering. Ljubljana, Slovenia, 2009.

50. Serope Kalpakjian; Steven R. Schmid. Manufacturing Engineering and Technology. International Edition. 4th Ed. Prentice Hall, Inc. 2001.

51. Tomblin, J.; Seneviratne, W.; Kim, H.; Lee, J. Characterization of In-Plane, Shear-Loaded Adhesive Lap Joints, Experiments and Analysis. Office of Aviation Research, USA[Online] 2003. http://www.tc.faa.gov/its/worldpac/techrpt/ar03–21.pdf (accessed March 5, 2015).

52. Walters, R. N.; Lyon, R. E. Flammability of Polymer Composites. Office of Aviation Research and Development, USA [Online] 2008. http://www.tc.faa.gov/its/worldpac/techrpt/ar0818.pdf (accessed Jan. 5, 2016).

53. Wróbel, G.; Pawlak, S. A comparison study of the pulse-echo and through-transmission ultrasonics in glass/epoxy composites. JAMME 2007, 22(2), 51–54.
54. NASA. Ultrasonic testing of aerospace materials. [Online] http://llis.nasa.gov/lesson/765 (accessed Jan. 5, 2016).
55. Hsu, D. K. Non nondestructive inspection of composite structures: methods and practice. 17th World Conference on Nondestructive Testing, Shanghai, China, 2008.
56. Kapdia, A. Best practice guide, non-destructive testing of composite material. National composite network. [Online] https://compositesuk.co.uk/system/files/documents/ndtofcomposites.pdf (accessed Aug 20, 2015).
57. Srinivas, K.; Siddiqui, A. O.; Lahiri, J. Thermographic Inspection of Composite Materials. Proceedings National Seminar on Non-Destructive Evaluation, Hyderabad, India, 2006.
58. Yang, L. X.; Hung, Y. Y. Digital stereography for nondestructive evaluation and application in automotive and aerospace industries. [Online], http://www.ndt.net/article/wcndt2004/pdf/optical_techniques/534_yang.pdf (accessed Sept. 10, 2015).
59. Hung, M. Y. Y. Shearography: A Full-Field Measurement Technique and Application in Non-Destructive Testing. SAE Technical Paper Series, No. 2005-01-0485, 2005. SAE World Congress, Detroit, Michigan, 2005.

INDEX

Printed in the United States
by Baker & Taylor Publisher Services